AWESOME Physics Experiments for Kids

AWESOME PHYSICS EXPERIMENTS for Kids

40 FUN SCIENCE PROJECTS & Why They Work!

ERICA L. COLÓN, PHD

Photography by Paige Green

SAFETY NOTE: All of the experiments in this book are intended to be performed under adult supervision. Appropriate and reasonable caution is required at all times, and the recommendations in the activities in this book cannot replace common and sound judgment. Observe safety and caution at all times. The author and Publisher disclaim all liability for any damage, mishap, or injury that may occur from engaging in the activities featured in this book.

For general information on our other products and services or to obtain technical support, please contact our Customer Care Department within the United States at (866) 744-2665, or outside the United States at (510) 253-0500.

Rockridge Press publishes its books in a variety of electronic and print formats. Some content that appears in print may not be available in electronic books, and vice versa.

Interior and Cover Designer: William Mack
Art Manager: Karen Beard
Editor: Brian Hurley
Production Editor: Erum Khan

Photography © 2018 Paige Green
Illustrations © 2018 Conor Buckley
Author Photo © 2018 Hannah Aleece Photography

ISBN: Print 978-1-64152-298-4
eBook 978-1-64152-299-1

To my junior scientists,
AVA, DANI, and LINCOLN. I love
watching how you learn, seeing
what sparks your creativity, then
exploring what excites you without
doubt or hesitation.

Contents

Introduction

By choosing *Awesome Physics Experiments for Kids*, I know that you and the young person in your life are curious because you are looking to understand how science shapes the world around us. Through these hands-on and fun experiments, you are both going to learn more about physics—the scientific discipline that seeks to understand the fundamental principles behind how the universe behaves. I started my career as a science teacher with the intention of teaching only biology. I laugh at that now because I've taught every science subject out there. For my first job, I was offered the opportunity to teach a course called Physical Science, which integrates concepts from both physics and chemistry.

Let me tell you, it was a game changer. Right from the beginning, when I started researching and discovering all the ways to approach the course, I got really excited! I knew my students were going to be telling their families about all the fun they had in my classroom. The best part was that I could teach almost everything I wanted with everyday supplies, easy-to-get items, and even recycled materials brought in from my students' homes. This led to earning my doctorate in the area of Curriculum and Instruction.

I wanted to share my student-approved, curriculum-based lessons with every science teacher and child I could, so I founded my company, Nitty Gritty Science. Now I'm helping teachers and students all over the world learn more about the nuts and bolts of science, and having a blast while doing so. Despite my love of all science disciplines, physics is near and dear to my heart. In this book, I'm sharing 40 of my favorite physics investigations and experiments for kids ages 10 to 13. But grown-ups can join in the fun, too, especially on activities marked with a safety reminder for your young scientists—look for the design element that resembles an exclamation mark.

If you ask a kid if they've studied physics, they will probably say no. But have they ever stared at a rainbow and wondered how it's made;

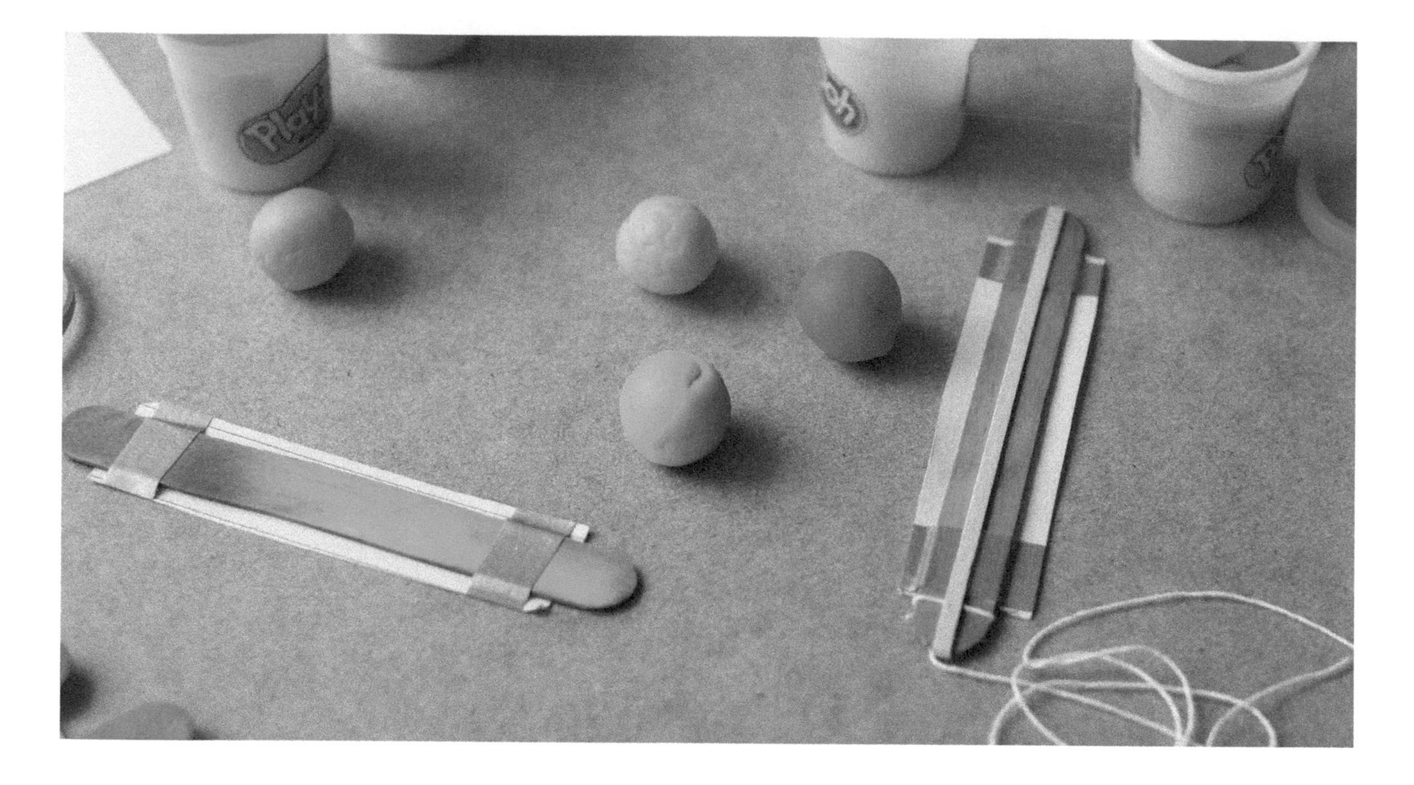

stood in an auditorium and hollered, "Hello!" over and over just to hear the sound of their echo; pushed a friend on a swing; or built a ramp to ride their bicycle off of? Kids, unknowingly, have been studying physics for as long as they can remember. Physics is happening not only all around us; its properties are happening to us every second of every day.

Physics comes from the Latin word *physica*, meaning "natural things." It is all about discovering how things like energy, motion, and force work. The investigations in this book will provide kids with meaningful, hands-on experiences that will inform them about the properties of the world around them. You'll see the moment they understand how and why something is happening—educators call that the "aha" moment. Those goosebump-raising experiences are what we live for in our profession.

Kids today are growing up in a world with so much complex technology and innovation that scientific literacy is key for them to make sense of it all. Scientific literacy begins with reading and using science terms, many of which are introduced in this book and can be found in the Glossary (page 108). I feel it is important that activities are age appropriate, so each experiment in this book is aligned with the Next Generation Science Standards, a national academic framework that outlines scientific concepts by grade level.

Awesome Physics Experiments for Kids will allow the budding scientist in your life to learn new information by performing investigations using an organized procedure called the scientific method. A scientist might add new steps, repeat some steps multiple times, or skip steps altogether when doing an investigation, but for this book, we will generally use this series of procedures of the scientific method:

- State the problem or question
- Gather information
- Form a hypothesis
- Test the hypothesis
- Collect and analyze the data
- Draw a conclusion
- Communicate the findings

This book is interactive. For each activity, there will be a proposed question or problem. Depending on what they already know, kids will gather information and form a hypothesis, or an educated guess, about what will occur in the investigation. Steps are laid out so that their hypothesis can be tested. During the experiment, kids should practice collecting and recording data. This can be done by noting observations, drawing pictures, or taking measurements. A careful observer then can analyze the data to draw a conclusion. The final step, communicating the findings, is super important since this is how one explains if the hypothesis was correct. If it was, great! If not, that's great, too! That's all part of thinking like a scientist. If

something didn't work, they need to use science skills to explain why their hypothesis wasn't correct and determine how to test it again. When the investigation is complete, kids can challenge themselves using the Now Try This! feature at the end of each activity.

Physics has such an important impact on our everyday life. Computers, planes, architecture, space travel, and even cooking are just some of the things that wouldn't be possible without the role physics-based ideas played in their development. It's important to know that the benefits of understanding physics will serve your child inside and outside the classroom. Your curiosity about how things work in the natural world tells me you and the young person in your life are going to enjoy this book and the awesome experiments that await you!

Chapter One

HOW TO USE THIS BOOK

I know you're ready to dive into the world of physics! However, there are a few things I would like to share before we begin. First, this book is divided into chapters by the following physics categories: Force (chapter 2), Buoyancy and Flight (chapter 3), Solids and Liquids (chapter 4), Electricity (chapter 5), Magnetism (chapter 6), Light (chapter 7), Heat (chapter 8), and Sound (chapter 9). Sometimes you might be doing an activity and think that it could have been in another chapter. You are probably correct, because many science concepts are connected. Sound is a perfect example: It is our brain's interpretation of a certain type of wave traveling at a certain speed through a certain medium, which can be either a solid, a liquid, or a gas. This interconnected nature is what's cool about physics. Once you start understanding one concept, you can apply it to another concept to help it all make sense! If an activity does happen to bridge categories, a design element at the top of the page will show you where else you could categorize the experiment. This first chapter will help us prepare for the great science experiments in the book, so let's get started!

GETTING READY

One of the great things about this book is that you can start on whichever activity you want—just close your eyes, open to a page, and get started. If you would like a little more to go on than chance, then you can start out by checking the Contents and find an experiment that catches your eye. It's important to make sure that both difficulty level and length of time needed are appropriate. Also, some experiments are messy and may need to be performed outdoors, or sometimes help with tools may be necessary, so check to see if an activity includes a !, which indicates you'll need adult help . . . or a ☼, which means it should be done outside. The ✹ means it could be a messy activity. Please read the directions for all of these activities and ask an adult for supervision to ensure your safety. You'll also see icons like or in the corner of some experiments (you should recognize them from the beginning of each chapter). This indicates the experiment demonstrates another element of physics in addition to the focus of that particular chapter.

Once you have decided on an experiment, carefully read through the materials list. Begin by collecting the items you need, which will usually be basic things you can find around your house. Adult help may be needed if you have to pick up an item from a local grocery store, or if you have to brainstorm a reasonable substitution. Next, you will need to find an appropriate place to conduct your investigations. A clean and organized work area is usually best so you know where everything is once you begin. It's important to remember that a good scientist will always leave the finished work area as clean as when they started.

It's always good practice to have a notebook and writing utensil handy when doing experiments. Not only can you write down questions and hypotheses, but you can also make drawings of observations, record notes, or collect measurements throughout the investigations.

Recording Observations: It's important that you record your observations in a notebook while conducting these experiments. This will help you keep track of whether something worked the way you thought it would, and why or why not. You can even decorate the cover of your notebook with stickers, drawings, and whatever else you like, or you can use colored pencils to record all of your findings. Keeping detailed notes will help you as you continue your quest for scientific knowledge.

I've started you off on each investigation with a "big question," which is a question you are trying to answer. If you have other questions related to the same topic, write them down in your notebook. Once you have familiarized yourself with the experiment, it's time to form a hypothesis. The hypothesis should be written before you begin your experimental procedures. It should be an educated guess on how or why something occurs, made as the following statement: "If [I do this], then [this will happen]." For example, if you are testing sunlight's effect on plants, a hypothesis may be, "If I wrap half the leaves in aluminum foil, then the wrapped leaves will turn brown." Once a hypothesis is stated, there's only one thing left to do: prove (or disprove) it!

A Note on Metrics: As you go through the activities in this book, one detail you may notice is that many measurements follow the metric system instead of the English system, which is the basis for what is commonly used in the United States. The reason for this is that the metric system is based on multiples of 10 and is easier to convert between different units. It is universally accepted and understood by scientists throughout the world, so what better time to start using it than now?

DOING THE EXPERIMENT

Now for the fun part! Scientists learn new information about the natural world by performing investigations. Some involve watching something that occurs and recording observations. Others involve setting up experiments that test one variable against another. Scientists also investigate phenomena by building a model that resembles something in the natural world, then testing it to see what happens. This book includes all three types of investigations. All have easy-to-follow instructions so you can complete the steps on your own, though some require adult supervision. When conducting the experiments, make sure to follow the steps carefully. By doing so, you will ensure consistent results. However, if you do want to make a variation to a step, write it down in your science notebook so you can share any new information or repeat the exact same steps if necessary.

Once you have completed the activity, use your observations, drawings, or data to go back and see if you can answer the big question. Are you able to make a conclusion and determine if your hypothesis was correct? What surprised you during the experiment? Are you able to explain what happened, or are you still a little unsure? It's okay to be unsure. If everybody knew everything about everything, then life would be pretty boring. Try the experiment again and keep observing. This is how you learn.

After you've completed an investigation, I encourage you to take it a step further and check out the Now Try This! section. This section challenges you to elaborate on what you know and apply it to a modified experiment or design your own investigation.

To help you understand the science behind the results, I've added a section after each investigation called The Hows and Whys. In this section, I explain the science happening in the activity, which may be exactly what you were thinking or may help you build new knowledge. This will only help you more when you move on to another experiment!

When you're learning science, you are constantly encountering new words and terms. If you are unsure of a word that appears in these activities, make sure to use the Glossary (page 108) to help you understand the meaning. This is all a part of the research process and may lead you to gather more information on the concept.

It's always important to remember that a successful investigation does not always come out the way you predicted it would. If an experiment does not work the way it's supposed to, use it as a learning opportunity and try to discover why something did or did not occur. History has shown that time and time again, failed experiments have led to some amazing discoveries and innovations. Most important, science is about having fun!

Chapter Two

FORCE

When you play tug-of-war or push your friend on a swing, you are applying force. Force is the push or pull upon an object resulting from its interaction with another object. In these examples, it's obvious a force is applied because objects are moving. However, some forces aren't so noticeable. Do you ever wonder why you don't slip and fall down every time you walk across the floor? Or why you can't throw a Ping-Pong ball as far as a baseball even though it's lighter in weight? Or why you always fall down to the ground instead up to the sky?

This chapter will begin to answer some of these questions by using everyday items to show the force of a slingshot, the super strength of paper, and the motion of what I call a jitterbean. You don't know what that is? Guess you'll have to keep reading to find out.

While working through these activities, keep in mind that force is related to mass and acceleration. This means that the heavier an object is or the faster it is, the more force it will have. So, always think of your safety and the safety of those around you when shooting, swinging, bouncing, dropping, pushing, pulling, spinning, thrusting, and balancing objects.

Don't forget to have your science notebook with you to record observations, especially if you are changing the variables. It's good practice to do more than one trial when conducting an experiment. This helps you see if the outcome is consistent and not altered by human or equipment errors. If something does go differently than expected, use that experience as a learning opportunity. Find out where it went wrong, try to fix the issue, then run the experiment again. That's science!

PAPER POWER

What force can lock two notebooks together?

LEVEL OF DIFFICULTY: EASY
TIME SUGGESTION: 10 MINUTES

MATERIALS

- 2 notebooks with approximately the same number of pages

THE STEPS

1. On a flat surface, open the notebooks to the last page and overlap the back covers.
2. Take the next pages from the notebooks and overlap each one. Continue alternating pages from each notebook until you reach the front covers.
3. Ask a friend or family member to grab the binding of one book as you grab the other, and try to pull the notebooks apart.

Now Try This! Determine the minimum number of pages that needs to be interwoven to make the books difficult to pull apart. Find out if you get the same results using small notepads, or try using something with a different type of paper, like catalogs or magazines.

The Hows and Whys: At first glance, the surface of a piece of paper seems very smooth. However, if you were to look at it under a microscope, you would see dips and bumps from its fibers. When two pieces of paper touch each other, the bumps come in contact and form microwelds. Microwelds are a source of friction, which is a force that causes resistance when two surfaces try to slide over each other. The amount of friction between two sheets of paper is easy to overcome, so you can slide one sheet over another. In this activity, however, when you try to overcome the friction of dozens or hundreds of sheets of paper, the force of your pull is not large enough to break the microwelds, so the notebooks are not able to pull apart.

POM-POM SLINGSHOT

How far can you shoot a pom-pom with the energy generated from rubber bands?

LEVEL OF DIFFICULTY: MEDIUM

TIME SUGGESTION: 30 MINUTES

MATERIALS

- Paper towel tube
- Scissors
- Large pom-poms (or large marshmallows)
- Tape
- Hole punch
- Pencil or chopstick
- 2 thin rubber bands

THE STEPS

1. Cut a paper towel tube in half crosswise using scissors. Take one half and cut it lengthwise so it's a rectangle.
2. Roll up that half into a tighter tube narrow enough to fit inside the other half. Test that a pom-pom can sit on the opening without falling through the tube. Secure the narrow tube with tape.
3. Use a hole punch to make two holes opposite each other on one end of the narrow tube. The holes should be about 1.5 centimeters from the end.
4. Gently push a pencil through the two holes until it is centered. Be careful not to tear the cardboard.
5. On one end of the wider tube, cut two slits, each about 1 centimeter long, about a fingertip-width apart from each other. Repeat on the same end, opposite the first pair of slits, for a total of four slits.
6. Gently slide one rubber band around a pair of slits. Secure it with tape. Repeat on the other side.
7. Place the narrow tube inside the wide one so that the pencil is on the opposite end of the rubber bands. Stretch the rubber bands to hook around each end of the pencil.

CONTINUED

8. Load the slingshot by resting a pom-pom on top of the inner tube and pulling back on the pencil. Launch the pom-pom by releasing the pencil!

Now Try This! Grab a friend and set up a target field of cups with a point system. Make up a game to test your speed and accuracy.

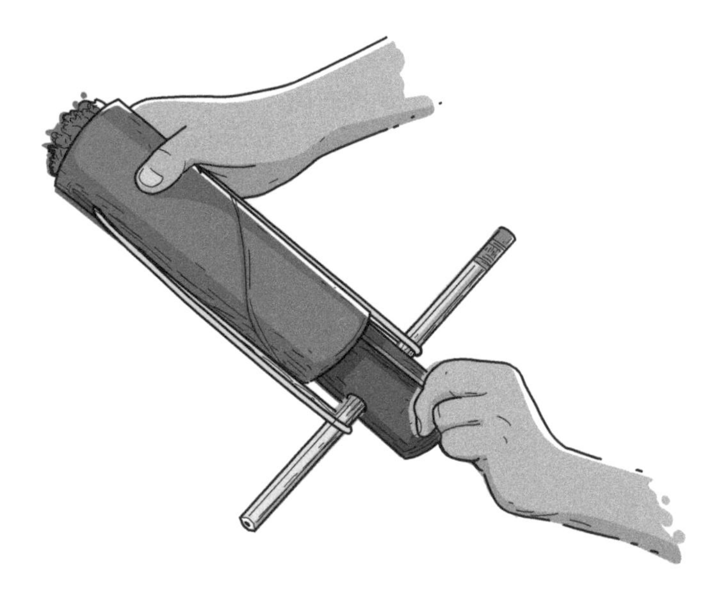

The Hows and Whys: Slingshots are made to shoot projectiles—a fancy term for objects that are shot or thrown—by using the stored elastic energy in rubber bands. When you release the rubber bands in this experiment, the stored energy transfers to the pom-pom and changes to kinetic energy—the energy of a moving object—which then sends the pom-pom flying through the air. To get the slingshot to fire the pom-pom at the greatest speed and distance, you will need to pull back as far on the rubber bands as you can without breaking them. This maximizes the stored energy in the rubber bands, which will give you the greatest kinetic energy when you release them.

TRIC-KEY

How can you stop keys from dropping on the floor when your eyes are closed?

LEVEL OF DIFFICULTY: EASY

TIME SUGGESTION: 10 MINUTES

MATERIALS

- 3 metal keys
- 1-meter-long string
- Large washer

THE STEPS

1. Securely tie the keys together to one end of the string. Tie the washer to the other end of the string.
2. Hold one arm out in front of you and point your finger. Place the string over your finger so that the keys are dangling 2 to 3 centimeters below it.
3. Holding the washer in your other hand, straighten out your arm so that the string is horizontal in front of you.
4. Close your eyes and let go of the washer to send the keys dropping. Did they hit the floor? Now you may want to repeat the experiment, this time with your eyes open!

Now Try This! Experiment with changing the angle of the string before you let go of the washer. Find the mass of the keys (mass = volume x density) and the washer and figure out the ratio. Can you find other household items with the same mass ratio and see if the experiment still works?

The Hows and Whys: The keys suspended on a string from a pivot, or fixed point, form a pendulum. Pendulums use acceleration from gravity and usually have a decrease in velocity when they swing. However, in this activity, the pendulum increases in velocity when you release the string, allowing it to make full swings and wrap around your finger. Eventually, the friction from the string around your finger stops the keys from hitting the ground.

JITTERBEAN

What causes a jitterbean to continuously dance about?

LEVEL OF DIFFICULTY: EASY

TIME SUGGESTION: 10 MINUTES

MATERIALS

- Aluminum foil
- Scissors
- Ruler
- Marker (with diameter slightly larger than a small marble)
- Small marble
- Small shoebox with lid

THE STEPS

1. Cut a piece of aluminum foil into a rectangle about 12 centimeters by 6 centimeters using scissors and a ruler.
2. Lay a marker on the short side of the aluminum foil and roll it up to make a 6-centimeter-long tube.
3. Slide the foil tube slightly off the marker and fold in the ends of the foil so that side is completely closed.
4. Carefully remove the foil tube from the marker and drop the marble inside. Close the open end of the foil tube so the marble cannot fall out but can still roll around inside the tube.

5. Place the foil tube inside a shoebox. Close the lid and shake the box side to side for 45 to 60 seconds. When done, check to make sure the tube now has round ends. You've just made a jitterbean!
6. Gently roll the bean around in the box or in your hand to see it jitter. Show your family and friends and see if they can figure out how it moves!

Now Try This! Try rolling the jitterbean along different surfaces and determine which is the best surface for making it dance. Try making some with larger marbles and see if that makes a difference in how the jitterbeans move.

The Hows and Whys: The jitterbean looks almost alive as it jitters about. This is caused by the rolling marble inside the foil tube. On certain surfaces, when the marble gets to the end, the tube will flip, causing the marble to roll again, keeping the jitterbean in constant motion. The marble will continue to roll around in the tube until a force stops it. This is a property of inertia, in which an object tends to resist any change in its motion.

NEWTON'S NOZZLE

How does a straw's angle affect the way a balloon spins?

LEVEL OF DIFFICULTY: EASY

TIME SUGGESTION: 15 MINUTES

MATERIALS

- Balloon
- Scissors
- Flexible drinking straw
- Tape
- Straight pin
- Pencil with eraser

THE STEPS

1. Blow up a balloon a few times so that it's stretched out and easy to inflate later in the experiment.
2. Use scissors to trim off the rubber ring at the balloon's opening.
3. Insert the long end of a flexible straw into the balloon's opening and secure the balloon to the straw with tape tight enough that no air can escape. Try testing for leaks by temporarily blowing up the balloon through the straw.
4. Find the straw's center of gravity by moving it until it can balance on one finger. At the straw's center of gravity, push a straight pin all the way through.

CONTINUED

5. Push the sharp end of the pin into the bottom of the pencil eraser, then stand the pencil on its tip on a flat surface.
6. Inflate the balloon through the straw and plug the end with a fingertip.
7. Bend the straw into a right angle parallel to the surface and let it go!

Now Try This! Try using two straws. Does this change how fast the balloon spins? What happens if you use two straws but bend them in opposite directions?

The Hows and Whys: A famous scientist by the name of Sir Isaac Newton stated that for every action, there is an equal and opposite reaction. This is the basic principle you see when a rocket is launched into space. The rocket engine burns fuel, producing hot gases that escape out an opening in the back of the rocket called the exhaust nozzle. These hot gases exert a force on the rocket, pushing it in the opposite direction. In this activity, when the balloon deflates, it pushes air out of the exhaust nozzle (end of the straw) at a right angle, which pushes the straw and balloon in the opposite direction.

MOMENTUM MAYHEM

How can you get a ball to bounce super high?

LEVEL OF DIFFICULTY: EASY

TIME SUGGESTION: 10 MINUTES

MATERIALS

- Variety of balls of your choosing, like a basketball, a tennis ball, a golf ball, and a Ping-Pong ball
- Chalk
- Tape measure

THE STEPS

1. Drop each ball from shoulder height next to a wall. Have a helper use chalk to mark how high each ball bounced. Measure the height with a tape measure.

2. Choose two balls—for example, a basketball and a tennis ball. Gently balance the tennis ball on top of the basketball. This part is a little tricky and will take some practice.

3. Hold the basketball out in front of you and drop it. What happens to the tennis ball? How does its height compare with the chalk marks?

4. Repeat steps 2 and 3 using a different combination of balls. Do you see any difference?

Now Try This! Try balancing a triple stack! Here's a hint: Use a golf ball in the middle since the dimples will help it stay balanced. Here's another hint: Get a friend to help hold while you balance.

The Hows and Whys: Moving objects have a property called momentum. Momentum of an object depends on its mass and velocity. Momentum doesn't change unless the mass, velocity, or both change. However, momentum can be transferred from one object to another. When you start by holding a basketball and a tennis ball, each ball has potential energy; when you drop them together, the basketball hits the ground and compresses, storing elastic potential energy. As it comes up, it acts like a trampoline for the tennis ball, not only transferring its energy but also transferring its momentum, sending the tennis ball flying super high.

MARBLE MANIAC

How can you adjust a ramp to make a marble jump the greatest number of cars?

LEVEL OF DIFFICULTY: EASY

TIME SUGGESTION: 30 MINUTES

MATERIALS

- Pool noodle
- Scissors
- Several large boxes (or a desk)
- Tape
- Ruler
- Play-Doh
- Large marble
- Several toy cars

THE STEPS

1. Ask an adult to cut a pool noodle in half lengthwise using scissors.
2. Stack the boxes. Take one half of the pool noodle and attach one end, cut-side up, to the stack of boxes using tape. This will be the starting point for the marble.
3. Find the spot on the ground directly below the beginning of the track. Use the ruler to measure the vertical distance between this spot and the beginning of the track. This is called the track's rise.
4. Curl up the other end of the noodle slightly and support it on a base made out of Play-Doh. The horizontal distance between the same spot on the ground directly below the beginning of the track and the lowest point of the track is called the track's run.

5. Place the toy cars at the end of the track for the marble to jump over. Position the marble at the top of the track and let it roll down. Observe how many cars it clears.
6. Adjust the rise length (maybe by adding another box to the stack) while keeping the run the same. Roll the marble down the track again and observe how many cars it jumps.
7. Then adjust the run by cutting the length of the pool noodle, while keeping the rise the same. Roll the marble down the track and observe how many cars it jumps.

Now Try This! Try launching marbles of different sizes down the ramp to see if they still jump the same number of cars. You can also try to reduce friction by adding a small amount of lubricant (like baby oil or olive oil) to the ramp to see if this affects the marble's speed and distance.

The Hows and Whys: The marble has potential (or stored) energy at the top of the track. As it rolls down, the potential energy is converted into kinetic (or moving) energy. If a marble has enough kinetic energy, it will successfully jump the cars at the end of the track. Know that the higher you place the marble to start, the more kinetic energy it will have when it reaches the bottom. Therefore, you should see little difference in jump length if you adjust the run, and greater difference in jump length when you adjust the rise.

SPECIAL DELIVERY

How does water stay in a cup when it's upside down and over your head?

LEVEL OF DIFFICULTY: MEDIUM
TIME SUGGESTION: 30 MINUTES

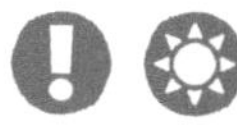

MATERIALS

- Empty pizza box (or 2 pieces of sturdy cardboard cut into equal squares)
- Scissors
- 2 (2-meter-long) heavy strings or ropes
- Duct tape
- Plastic cups
- Water
- Food coloring (optional)

THE STEPS

1. Cut the sides off a pizza box using scissors so you are left with the top and bottom of the box.
2. Stack the top and bottom of the box and tape together along the edges with duct tape to create a sturdy surface.
3. Lay the strings in a large *X* on the cardboard so that the center of the *X* is at the center of the cardboard and the ends of the strings are stretched out over the corners.

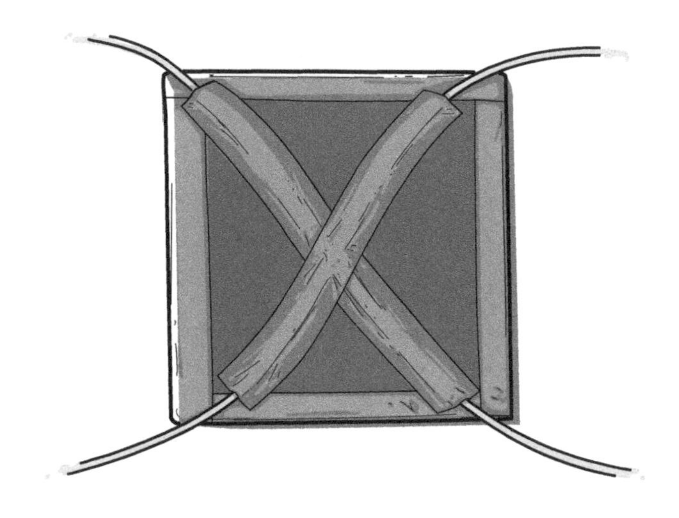

4. Use duct tape to securely fasten the strings to the pizza box, making sure to press down all the way from the center out to the corners.
5. Flip the cardboard over, bring the ends of the strings together, and make a knot. Hold up the strings at the knot to make sure they are even and the cardboard lies flat.
6. Find an outdoor area with enough room to swing the box around you without hitting anything. If you need to, tie a knot further down.

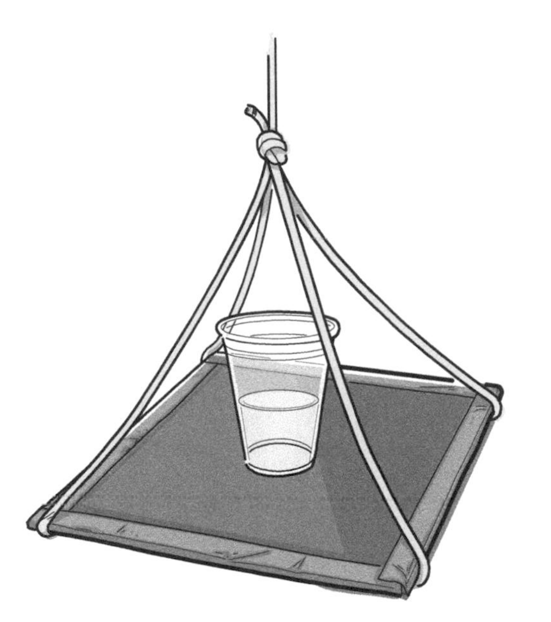

7. Place a plastic cup filled with water and food coloring (if using) in the center of the cardboard. Tell your thirsty friend you'll bring them a drink.
8. Slowly begin swinging the cardboard back and forth in an arc; then, when you are ready, quickly swing it in a complete circle above your head! Gradually bring it to a stop and deliver the water to your friend, who will have their mouth hanging open in awe.

Now Try This! Once you've mastered this activity, try adding more plastic cups filled with water. Do they all have to be on the same level, or can you stack them in a pyramid with cardboard in between the layers?

The Hows and Whys: The water, cup, and cardboard have inertia, meaning that once in motion, they tend to remain in motion unless acted upon by another force. The pull of the string keeps the cardboard and everything on it moving in a circle rather than a straight line. The force of the circular motion, called centripetal force, is greater than the force of gravity on the water. As long as you're spinning it fast enough, the water won't spill all over your head.

Chapter Three

BUOYANCY AND FLIGHT

It's amazing how planes fly and aircraft carriers float, but many people don't stop to wonder how these great feats are even possible. The physics behind buoyancy and flight deals with properties of things that flow, such as liquid and gas. These properties will help you in the following activities as you discover how to make things fly, float, spin, and sink. In this chapter, you will find that not only do these properties determine the design of ships and airplanes, but they also are a factor when playing sports and even when inventing better life jackets.

One of the main properties you will explore is buoyancy, which is the ability of a fluid to push up on an object immersed in it. Of course, you'll be using one of the most abundant fluids there is: water. So, make sure to do these experiments where it's okay to get wet and where cleaning up will be easy.

You will also be giving flight to some creations using air pressure. If the experiments don't work well the first time, I would advise you to look at your structures. Did you fold, cut, or bend in nice, straight lines? Is there anything adding unnecessary weight, like too much tape? Many times, the failure is caused by factors that you are able to easily correct in order to have a successful outcome. Now go make Daniel Bernoulli proud (you know, the scientist and mathematician who discovered the main properties of fluids), and go float an egg!

EGG-CITED ABOUT DENSITY

How can you suspend an egg in the middle of a glass of water?

LEVEL OF DIFFICULTY: EASY

TIME SUGGESTION: 10 MINUTES

MATERIALS

- 3 clear, tall drinking glasses
- Water
- Dry-erase marker
- Salt
- 3 eggs

THE STEPS

1. Fill each glass with 250 milliliters of water and label them A, B, and C with the marker. Add 4 teaspoons of salt each to glass B and glass C. Stir the contents of each glass until the salt dissolves.
2. Carefully add an egg to each glass and record your observations.
3. Slowly add fresh water to glass C until it is almost full. Record your observations.

Now Try This! Set up an experiment with other solutions to see how their densities compare with the density of eggs. For example, you could try using sugar instead of salt to see if you get the same results. Another option would be to hard-boil the eggs first and see if you get the same results.

The Hows and Whys: Saltwater is denser than tap water. This means that there is more stuff (matter) in the same amount (volume) of water. If an object has a density that is lower than the density of the water it is placed in, it will float. In this activity, an egg has greater density than tap water, so it sinks (glass A) but has lower density than more concentrated salt water, so it floats to the top (glass B). In glass C, the saltwater is diluted enough that the saltwater and the egg have the same density, allowing the egg to float in the middle of the cup.

THROW 'EM A CURVEBALL

How do baseballs, soccer balls, and golf balls curve when hit or thrown?

LEVEL OF DIFFICULTY: MEDIUM

TIME SUGGESTION: 15 MINUTES

MATERIALS

- Tape
- 2 disposable cups
- Markers
- 4 rubber bands

THE STEPS

1. Securely tape the bottoms of the cups together to make a dumbbell shape. This is your flier. Decorate your flier with markers.
2. Loop the rubber bands together by sliding one through another and then back through itself until you have a chain.
3. Firmly holding one end of the rubber band chain at the center of the flier, wrap it tightly around the center about four times.

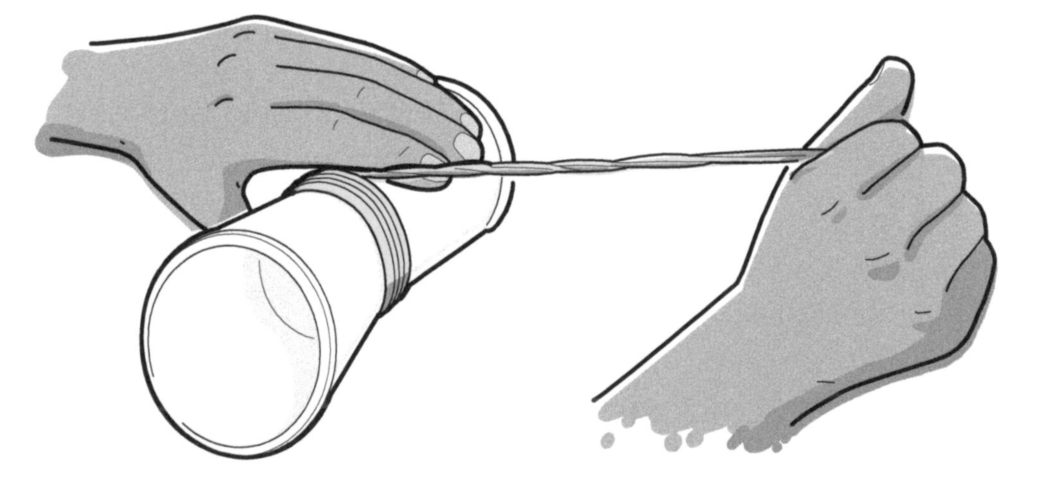

4. Prepare for launch by holding the end of the rubber band chain stretched out on the underside of the flier and the other hand pulling back on the flier by the center.

5. Now launch it. Let go and observe how it flies.

Now Try This! There are a lot of variables you could switch up in this experiment—just remember to write them down in your science notebook. Try using a variety of cup sizes or movie popcorn buckets to see if they have an effect on the flight path. What happens if you stretch out the rubber band on the upper side of the flier and let go? What happens if you launch your flier vertically?

The Hows and Whys: The cause of the upward path of this curvy flier is due to the Magnus effect, well known among many baseball pitchers. When a spinning cylinder moves through the air, it will create a pressure difference between the two sides. To demonstrate this, blow across the top of a sheet of paper and watch it rise. The speed of the air you blew over the top of the paper is greater than that of the slow air below it. As a result, the pressure on top is lower than the pressure underneath, allowing it to rise. With your curvy flier, the air on the side that is spinning in the opposite direction of the flier creates a low-pressure area, causing the upward curve.

ROTO-MOTO

What effect does the number, size, or shape of blades have on the flight of a rotor blade flier?

LEVEL OF DIFFICULTY: MEDIUM

TIME SUGGESTION: 20 MINUTES

MATERIALS

- Tracing paper
- Ruler
- Pencil
- Scissors
- Marker
- Card stock
- Book

THE STEPS

1. Use tracing paper, a ruler, and a pencil to trace the rotor patterns on page 32. Cut out the patterns using scissors and use them as a template.
2. Use a marker to trace the patterns onto a piece of card stock and cut them out.
3. Choose one rotor and place it on a book with one blade extending over the edge.
4. Tilt the book slightly upward and firmly hit the extended blade with your finger to start it spinning. Observe how it flies.
5. Repeat steps 3 and 4 with the other rotor. Compare your observations.

CONTINUED

ROTO-MOTO, CONTINUED

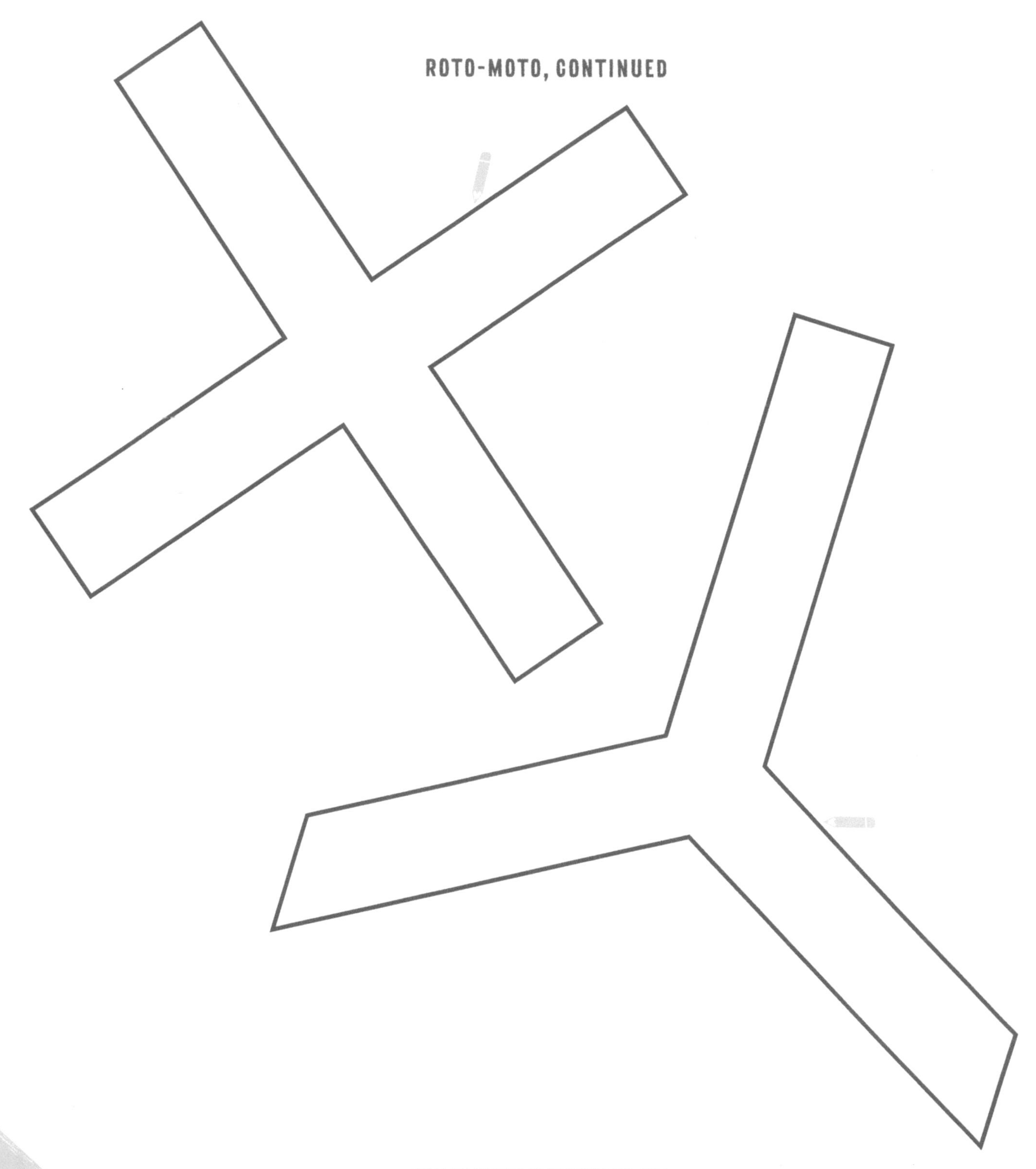

Now Try This! Design your own rotor that has a different size, shape, or number of blades. Is there a difference if you spin it clockwise versus counterclockwise? If you use a different weight of paper, does that make a difference?

The Hows and Whys: Rotor blades are the wings on helicopters that produce lift by rotating through the air. This is the same way lift is produced by airplane wings. Fast-moving air over the top of the rotor blades creates low pressure, while the air below the blade is slower, creating higher pressure. Higher pressure under the rotor blades creates lift, causing them to rise.

ROCKY'S LIFE VEST

Can you design and construct a life vest to keep a large rock afloat?

LEVEL OF DIFFICULTY: MEDIUM
TIME SUGGESTION: 1 HOUR

MATERIALS

- Large rock (about the size of your fist)
- Household materials to construct and attach your life vest, like string, packing peanuts, corks, paper towels, balloons, glue, aluminum foil, small containers, and straws
- Fish tank (or plastic bin)
- Water

THE STEPS

1. Consider the available materials and design a life vest for your large rock, knowing the following conditions: The life vest must attach as one piece, it must attach in 10 seconds or less, and it must keep the rock floating in the water, meaning that part of the rock must be submerged.
2. Once you have finalized your design, build the life vest. If you make changes to your design, be sure to mark the changes in your science notebook.
3. Test your rock and its life vest by placing them in a fish tank filled with water. Imagine "Rocky's" life is at stake. He's counting on your design to save him.

Now Try This! Challenge your friends and set up a time limit in which to build the best life vest. Make it more fun by personifying your rocks by painting some outfits, hair, and faces on them.

The Hows and Whys: When placed in water, Rocky will sink because the weight of the rock is greater than the buoyant force of the water. Therefore, Rocky requires a life vest. The first life vests, which have been improved over time to prevent the loss of life for humans, consisted of cork and wood, later being replaced by foam. Today, life vests use pockets of air, or bladders, that can be inflated either by a chemical reaction or by blowing into a tube.

STACKING LIQUIDS

What is the effect of density and buoyancy on different objects?

LEVEL OF DIFFICULTY: EASY

TIME SUGGESTION: 15 MINUTES

MATERIALS

- Corn syrup
- 100-milliliter beaker (or clear glass)
- Graduated cylinder (or liquid measuring cup)
- Food coloring
- Water
- Vegetable oil
- Aluminum foil crumpled to a ball the size of a pea
- Steel nut
- Whole peppercorn

THE STEPS

1. Pour 10 milliliters of corn syrup into a beaker.
2. In a graduated cylinder, combine a few drops of food coloring with 10 milliliters of water and stir. Slowly pour the mixture into the beaker.
3. Next, add 10 milliliters of vegetable oil to the beaker.
4. Carefully drop the aluminum foil ball, steel nut, and whole peppercorn into the beaker. Observe where each object settles.

Now Try This! Try using other liquids with different densities, such as liquid soap, glycerin, colored salt water, and so on. Then try adding other objects to test their buoyancy, such as a cork, an eraser, or a bean.

The Hows and Whys: The corn syrup has the highest density, or the greatest mass per unit volume; therefore it sinks to the bottom. The water has less density than the corn syrup, but more than the vegetable oil, which stays on top. The weight of the foil is less than the buoyant force (ability of a liquid to exert an upward force) of the displaced oil, so it floats on top of the oil. The steel nut sinks to the bottom since its weight is greater than the buoyant force of any of the liquids. The peppercorn's weight is greater than the buoyant force of the oil but less than that of the displaced water. That's why it settles in the middle.

Chapter Four

SOLIDS AND LIQUIDS

States of matter are like the crowd at a concert. When seated, people are like particles in a solid—they can move in place but don't go anywhere. The people in the aisles are like liquid particles—they can easily pass one another but aren't free to move very far apart. So what state of matter are people like once the concert is over and the doors of the concert hall open up to let them freely go home? You guessed it: gas.

All three states of matter can be seen all around us, but for this particular chapter, the activities focus on solids and liquids since we are able to clearly see the properties of these states of matter (although I did sneak some gas in there, too)!

I'm warning you, as you explore the following experiments, you will be challenged on what you already know about solids and liquids. Get ready for some unexpected changes, some unforeseen outcomes, and maybe even some astonished shrieks of surprise from you or your audience!

One particular liquid we will use in the experiments has very unique properties. Want to guess what it is? Water! Because of the shape of water molecules, they almost break the rules of how liquids should act. After you discover some of these unusual properties in the upcoming activities, I encourage you to research water and learn even more. You'll be amazed! Some solids also have unusual behavior. They are known as amorphous solids, from the Greek word for "without form." Two amorphous solids you are familiar with are glass and plastic. But what makes them so unusual? Get ready to find out!

MINI SNACK BAG

What happens when you heat an empty chip bag?

LEVEL OF DIFFICULTY: EASY

TIME SUGGESTION: 20 MINUTES

MATERIALS

- 1-ounce bag of chips
- Baking sheet
- Parchment paper (or aluminum foil)
- Oven mitts

THE STEPS

1. Ask an adult to help you preheat the oven to 250°F.
2. While waiting for oven to heat up, enjoy the chips. Wash and dry the empty chip bag.
3. Line a baking sheet with parchment paper. Lay the chip bag flat on the parchment paper and place the baking sheet in the oven on the middle rack for 10 minutes. Ask for an adult's help if you need it.
4. Have an adult remove the cookie sheet using oven mitts and let it cool for a few minutes.
5. Carefully peel the parchment paper away from your mini chip bag.

Now Try This! Shrink more chip bags and use them to make jewelry, gift tags, play house food, or art projects.

The Hows and Whys: Chip bags are made of aluminum foil to keep the chips fresh, but each bag has a thin piece of plastic that covers the aluminum foil. The plastic is made up of a polymer, or a long chain of molecules that prefer to be in a knotted string. However, in order to be used for chip bags, the polymer is stretched flat over the aluminum. When the plastic is heated, the solid particles that make up the plastic start to vibrate faster and are able to revert from the stretched state back to their natural, bunched-up shape. Since the aluminum and paint on the plastic are still bound together, the overall bag shape is maintained when shrunk.

CAN COLLAPSE

How does a drop of water implode an aluminum can?

LEVEL OF DIFFICULTY: EASY

TIME SUGGESTION: 15 MINUTES

MATERIALS

- 1 tablespoon of water
- Empty aluminum can
- Hot plate (or electric stove burner)
- Bowl filled with ice water
- Tongs large enough to fit around a can

THE STEPS

1. Put the water in an empty aluminum can.
2. With adult supervision, heat the can on a hot plate until the water is boiling and steam is escaping from the opening.
3. Place a bowl of ice water next to the hot plate.
4. With adult supervision, use tongs to grab the can. Carefully and quickly flip the can upside down into the bowl of ice water. Boom! Did you shriek?
5. Repeat steps 1 through 4 with another can—I know you want to!

Now Try This! Try changing the conditions of the experiment, one variable at a time. An example would be instead of getting the can really hot, try putting it in the freezer. Or try placing the hot can in the ice bath right-side up.

The Hows and Whys: The steam coming out of the can is the liquid water you started with getting turned to water vapor when it was heated. Not only does the steam escape out of the can, but it also pushes out any air that was in there before you heated it. When the can is flipped over into the ice water bath, the remaining steam cools, condensing back into liquid water. The amount of steam left in the can is equal to about one drop of water. Since all the air was pushed out earlier and now there is only one drop of water left in the can, the air pressure on the outside of can pushes in on it, causing a sudden collapse, or implosion.

SNEAKY SURFACE

How does water defy the laws of gravity when turned upside down in an open jar?

LEVEL OF DIFFICULTY: EASY

TIME SUGGESTION: 20 MINUTES

MATERIALS

- Small mason jar with screw-on metal band
- 10-by-10-centimeter piece of screen
- Marker
- Scissors
- Water
- Large card
- Bowl
- Needle

THE STEPS

1. Place a mason jar upside down on the screen. Use a marker to trace the top of the jar. Using scissors, cut the circle out of the screen and place it on top of the jar. Hold the screen in place by screwing on the metal band.
2. Pour water into the mason jar through the screen until it's almost full.
3. Place a card flat on top of the jar and hold it in place while you flip the jar upside down over a bowl. (The bowl is there as a "safety net.")
4. Slowly slide the card off the opening of the jar while it's upside down. The water should stay in the jar.
5. Carefully push a needle through one of the openings in the screen. The needle should float to the top.
6. Replace the card over the opening, turn the jar right-side up, remove the card, and pour out the water! After practicing a few times, go amaze your friends and family!

Now Try This! Try different screens and see what is the largest-size mesh you can use to hold the water. If the mesh is bigger, also try pushing larger objects through, like a toothpick or skewer, to see if the water still stays in the jar.

The Hows and Whys: This experiment has a lot going on. First, atmospheric pressure, or the tiny air molecules around us, are pushing up on the card, holding it in place. There are more air molecules pushing up against the bottom of the card, creating a higher pressure area compared with the lower pressure area inside the air pocket in the jar you created when you flipped it upside down. The second part deals with one of water's coolest properties: surface tension. The surface of water acts like it has a thin skin because of a force called cohesion, or the attraction of water molecules to themselves. When the jar is flipped over, water molecules hold onto the screen (adhesion) as well as hold onto one another (cohesion) across the tiny openings in the screen, creating surface tension. Pushing the needle through the screen will add to the illusion that there is a "magic" layer keeping the water in.

EGG-STREME FOAM

How can you make instant foam from egg whites?

LEVEL OF DIFFICULTY: MEDIUM

TIME SUGGESTION: 20 MINUTES

MATERIALS

- Vitamin C tablets
- Small plastic bag
- Teaspoon
- 1 egg
- Small bowl
- Empty plastic water bottle
- Water
- Baking soda
- Clear plastic cup
- Paper plate

THE STEPS

1. Place a vitamin C tablet in a plastic bag. Use the back of a spoon to press down through the plastic bag firmly against a table to break the vitamin apart. Continue to do this, adding more vitamin C tablets one at a time if needed, until you have about 1 teaspoon of powder.

2. Separate the yolk from the egg white by using this cool pressure trick: Crack the egg into a bowl. Take an empty plastic water bottle and squeeze in the sides. While squeezing, place the opening of the water bottle over the yolk and slowly release the water bottle. Discard the yolk.

3. Place one teaspoon each of egg white, water, and baking soda into an empty clear plastic cup. Stir well to combine.
4. Add the vitamin C powder to the cup and swirl the contents of the cup vigorously for 12 seconds.
5. Quickly turn the cup upside down on a paper plate. Repeat the experiment with the remaining egg white.

Now Try This! Ask an adult to help you find a recipe for meringue cookies. Spend time together and make a batch. Use what you learned from this experiment to explain how these cookies are so light and airy. Try adding different flavors for another twist!

The Hows and Whys: When you dissolve the solid baking soda and vitamin C powder (citric acid) in water, they react to form carbon dioxide gas. Tiny bubbles produced by the gas are trapped in the egg white. When shaken, this bubble-filled liquid mixture reacts with the acid, transforming the makeup of the egg into a frothy structure that now also has properties of a solid! When something shares properties of a liquid, solid, or gas, it is called a colloid.

Chapter Five

ELECTRICITY

Electricity is something we use on a daily basis. Without it, the lights wouldn't work, the refrigerator wouldn't keep food cold, and all the devices that need charging would eventually stop working.

In this chapter, you will start to learn about electric charge. Positive electric charge comes from protons, and negative electric charge comes from electrons. Both are found in atoms that make up all matter around us. However, you will find it's the movement or transfer of electrons that causes electricity. The next time you shock your finger when you touch a doorknob, or your sock sticks to your pants when you pull them out of the dryer, blame those darn electrons.

The next activities will be shockingly fun, because even though we can't see electrons moving, we can see lights and other electrical devices work by converting electrical energy into other forms of energy. Using batteries, wires, switches, and circuit breakers, we are able to control the flow of electrons by giving them a circuit, or a path to follow. In order for electrical devices to work, a complete circuit must be made. This is especially important to remember when something isn't working correctly in an activity in this chapter. If something doesn't work, do an error analysis. This is what scientists do to check for possible sources of error. Is the battery dead? Did the bulb burn out? Are there poor connections where metal should be touching metal? Once the error is found, you'll now know to check for this in future experiments. Are you charged up for these next activities? Let's power up and go!

A FIVE-CENT FLASHLIGHT

How can you make a flashlight with pennies?

LEVEL OF DIFFICULTY: MEDIUM

TIME SUGGESTION: 20 MINUTES

MATERIALS

- 100-grit sandpaper
- 5 pennies (dated after 1982)
- Salt
- Cup filled with water
- 2 tablespoons of vinegar
- Cardboard, cut into 4 circles the size of pennies
- Paper towel
- LED light
- Clear or electrical tape

THE STEPS

1. Use sandpaper to remove the copper from the tails side of 4 of the pennies. You should sand until you see the zinc (shiny silver) over the entire side of the coin.
2. Add salt to the water a little at a time and stir. Keep adding salt until the solution is supersaturated (meaning no more salt dissolves). Then add the vinegar and stir to combine.
3. Put the cardboard pieces in the salt and vinegar solution until they are soaked through. Remove and lay on a paper towel.
4. Build the battery of the flashlight by taking one of the sanded pennies, zinc-side up, and placing a piece of wet cardboard on top. Stack another sanded penny (zinc-side up) and another wet cardboard piece. Continue with the rest of the sanded pennies and cardboard.

CONTINUED

5. Place the final, unsanded penny on top of your stack. Check to make sure that all zinc sides are facing up with a layer of cardboard covering each and that the top and bottom of the battery stack are copper. Note: Pennies should not be touching each other, and cardboard pieces should not be touching each other.

6. Turn on your flashlight by connecting an LED light to the penny battery. Do this by touching the longer leg of the light to the top penny (the positive end) and the shorter leg to the bottom penny. Once the light is lit, secure it with tape.

Note: If the light doesn't turn on, make sure no identical layers are touching and that all excess water has been wiped away.

Now Try This! Can you make a stronger battery by using more pennies? Can you recharge your battery by resoaking the cardboard pieces?

The Hows and Whys: Pennies made after 1982 use zinc cores plated with copper, making them perfect to build a battery. When the two different metal surfaces, electrodes, are connected by an electrolyte—the salty vinegar solution in the cardboard—a chemical reaction occurs. The zinc surface of the penny reacts with the saltwater to create electrons, and the copper surface of the penny reacts with the saltwater to use those electrons. When the LED is connected, electrons begin to flow, which lights the bulb.

SWITCH IT ON!

How does a switch turn a light bulb on and off?

LEVEL OF DIFFICULTY: MEDIUM

TIME SUGGESTION: 45 MINUTES

!

MATERIALS

- Broken strand of holiday lights
- Wire cutter
- 9-volt battery
- Cardboard cut out from an empty cereal box
- Marker
- Ruler
- Tape
- 10 (1.5-centimeter-wide) strips of aluminum foil
- 3 brass fasteners

THE STEPS

1. Ask an adult to help you cut a holiday light from the strand and expose a small section of copper at the end of each wire using a wire cutter. You should have a light bulb attached to a pair of wires, and each wire should have a small section of exposed copper. Test the light by touching one wire end to the positive terminal of 9-volt battery and the other end to the negative terminal of the battery.
2. Draw a 7-by-7-inch square on the cardboard using a marker and a ruler. On one side of the square, mark a 1-inch gap for the light bulb. On another side, mark a 1-inch gap for the switch. On a third side, mark a ¼-inch gap for the battery.
3. Tape down aluminum foil strips where you have drawn your square. Make sure to leave gaps in the aluminum foil where there are gaps in your drawing. Also make sure when you tape down the foil strips that each strip is touching another strip.
4. At the gap for the light, tape the light bulb down so that each wire is touching foil on either side of the gap. At the gap for the switch, push a brass fastener through the aluminum foil and cardboard on either side of the gap. Open the fasteners on the backside to keep them in place.

CONTINUED

5. Complete your switch by pushing the third brass fastener up through the center of the gap from the back of the cardboard. Open the fastener so that each "arm" can touch the head of the other two brass fasteners. When the center fastener touches the others, the circuit is closed, which will turn the light on. When the switch is turned so it's not touching, the circuit is open and will turn the light off.
6. Finally, place the battery on the ¼-inch gap so that the terminals are touching the aluminum foil on either side of the gap. Remember, metal must touch metal. While the battery is on the foil, turn your switch on and off.

Now Try This! Can you add more lights to your circuit? Can you add another switch and branches of aluminum foil to your circuit, then have certain switches control certain lights?

The Hows and Whys: When you connect the battery to the aluminum foil, electrons come out of the negative end of the battery and travel through the aluminum foil and the wires in the light bulb until they reach the switch. When the switch is on (metal touching metal), the electrons will continue to travel back to the positive terminal of the battery. This path is called a complete circuit. When the switch is off (not touching), the electrons cannot travel through and the circuit is incomplete, so the light stays unlit.

STATIC GOO

What makes goo attracted to a balloon?

LEVEL OF DIFFICULTY: EASY

TIME SUGGESTION: 20 MINUTES

MATERIALS

- 1/3 cup cornstarch
- 1/3 cup vegetable oil
- Bowl
- Balloon
- Spoon

THE STEPS

1. Mix the cornstarch and vegetable oil in a bowl until combined.
2. Blow up a balloon and secure it with a knot. Rub the balloon on your hair or the hair of a willing volunteer.
3. Use a spoon to pick up the gooey cornstarch mixture and let a stream drip off the spoon back into the bowl.
4. As the goo drips back into the bowl, place the balloon close to the stream of goo and record your observations.

Now Try This! Try turning on a thin stream of running water and see if you can get the same results with the static charge of the balloon.

The Hows and Whys: When you rub a balloon on your hair, some of the electrons from your hair transfer to the balloon, making the balloon more negatively charged. The cornstarch goo has a neutral charge, meaning it has the same number of protons and electrons. When the negatively charged balloon is near the goo, the electrons of the goo move away from, or are repelled by, the electrons in the balloon, and the protons of the goo move toward, or are attracted to, the balloon.

POPPING CIRCUIT BREAKERS

How does a circuit breaker work to keep wires from overheating?

LEVEL OF DIFFICULTY: DIFFICULT

TIME SUGGESTION: 20 MINUTES

MATERIALS

- Balloon
- 3 (3- to 5-centimeter-long) strands of tinsel (or steel wool)
- Tape
- 2 (30-centimeter-long) strips of aluminum foil
- 1 (7- to 10-centimeter-long) strip of aluminum foil
- Switch from Switch it On! (page 51)
- 2 to 3 D-cell batteries
- Electrical tape (if needed)

THE STEPS

1. Blow up a balloon and secure it with a knot.
2. Tape the middle of each strand of tinsel to the center of the balloon that so they are parallel to one another and a few millimeters apart.
3. Tape the long aluminum foil strips to the ends of the tinsel strands so that one foil strip covers all three strands on each end.
4. Secure one of these aluminum foil strips to one brass fastener of your switch. Make sure the switch is open.
5. Tape the other aluminum foil strip to the negative end of a D-cell battery.

6. Tape one end of the short strip of aluminum foil to the positive end of the battery, and secure the other end to the other brass fastener head of the switch.

7. Close the switch for 5 seconds. Record your observations. If nothing happened, try adding another D-cell battery to your electrical circuit. (Lay the batteries head to foot with the foil strip on each far end. Use electrical tape to hold them together, if needed.)

Now Try This! Can you change the amount of current carried by the circuit breaker before the balloon pops?

The Hows and Whys: As you learned in Switch it On! (page 51), a switch is used to open and close a circuit. But a switch needs to be manually turned off and on. One way to prevent wires from overheating in a closed circuit is by installing circuit breakers. Circuit breakers are designed to pop or interrupt the circuit when the current gets too hot. In reality, there aren't balloons that pop in your circuit breaker box in your home, but this activity models how a device can stop electricity flow.

Chapter Six

MAGNETISM

In 1820, a physics teacher named Hans Christian Oersted discovered that electricity and magnetism are related. Did you know that magnets are used in televisions, computers, speakers, electric motors, and smart-phones? Heck, even the Internet is dependent on magnets!

Depending on which way you hold magnets, they will either attract or repel each other, and you can feel this force before the magnets even touch. This is because magnets are surrounded by a magnetic field, which is stronger closer to the magnet and weaker farther away from the magnet.

This chapter challenges you to tinker around with magnets. By doing so, you'll perform amazing feats, such as turning common household items into magnets and creating an electric current to produce a magnetic field.

Before you start the activities in this chapter, it's very important to remember that magnets can be very dangerous for small children and pets if swallowed, particularly the small, high-powered ones often found in magnetic building sets and other toys. Therefore, it's always good practice to count items, including magnets, before you start any experiment, and make sure you have the same number when finished. This ensures nothing has been left where smaller children or pets can easily find them. Now that I know you'll be a safe and responsible scientist, let's go see what these magnets can do!

CONTROLLING ELECTROMAGNETS

How can you create a temporary magnet using an electric current?

LEVEL OF DIFFICULTY: EASY

TIME SUGGESTION: 20 MINUTES

!

MATERIALS

- 3-inch nail made of iron or zinc
- Metal paper clips
- Wire cutter (or scissors)
- Insulated copper wire
- Masking tape (or electrical tape)
- D-cell battery

THE STEPS

1. Attempt to pick up paper clips with the nail to confirm that it's not magnetized.
2. Ask an adult to help you use a wire cutter to expose the copper at the ends of a piece of insulated wire.
3. Wrap the insulated wire 10 times around one of the nails. Tape one end of the wire to the positive end of a D-cell battery and the other to the negative end of the battery. You have now made an electromagnet out of the nail.
4. Hold the electromagnet over the paper clips and pick up as many as possible. Count how many you picked up. Disconnect the wire from the battery and return all the paper clips to the pile.
5. Repeat the experiment, but this time wrap the wire 20 times around the nail. Record your observations.

CONTINUED

Now Try This! Try wrapping the wire 30 or 40 times around the nail and see if there is a difference in the number of paper clips you can pick up. Is there a difference if you wrap the wire neatly or if you wrap it messily around the nail? What if you add another battery using electrical tape?

The Hows and Whys: A magnetic field exists around any wire that carries an electric current. When the coil of wire wrapped around the nail is connected to the battery, a current begins to flow, making an electromagnet. An electromagnet is another type of temporary magnet, since it will work only if there is an electric current. By coiling the wire around the nail, you increase the strength of the magnetic field. The more coils there are, the stronger the electric current; therefore the magnetic force will also increase.

MAGNETIC SLIME

How can you make slime look like it's moving on its own?

LEVEL OF DIFFICULTY: EASY

TIME SUGGESTION: 20 MINUTES

MATERIALS

- 4-ounce bottle of white school glue
- 2 mixing bowls
- Water
- ½ teaspoon borax powder
- Spoon
- Iron filings
- Strong magnet (made out of neodymium)

THE STEPS

1. Empty the glue into one of the mixing bowls. Fill the empty glue bottle halfway with water. Close it and shake to release the rest of the glue. Add the glue water to the mixing bowl.
2. In the other bowl, add the borax powder to ¼ cup warm water. Stir until the borax dissolves. Then add the borax solution to the glue in the first bowl.
3. Mix the glue and borax solution together using a spoon, or just using your hands, until a large ball of slime is formed.

CONTINUED

4. Lay the slime on a flat surface and sprinkle 1 to 2 tablespoons of iron filings on top. Use your fingers to knead the iron filings into the slime until all the iron filings are spread throughout.
5. Bring the magnet close to the surface of the slime without touching it and see what happens. Record your observations.

Now Try This! Can you pull off a small section of slime and have it roll across the floor after the magnet?

The Hows and Whys: The iron filings are attracted to the strong magnet, but the adhesion between the iron filings and slime is also strong. Remember from chapter 4, Solids and Liquids, that adhesion is an invisible force that holds molecules of different substances together. The iron filings are caught in a tug-of-war between the magnet and the adhesive forces, which makes it look like the slime is reaching up and trying to grab the magnet.

MAGNETS IN MOTION

Can you create a scene with features that move by magnet power?

LEVEL OF DIFFICULTY: MEDIUM

TIME SUGGESTION: 1 HOUR

MATERIALS

- Cardboard from 2 cereal boxes or shoebox lids
- Materials to create a background and moveable objects or characters, such as markers, crayons, colored pencils, construction paper, pipe cleaners, and tissue paper
- Tape
- Scissors
- Metal paper clips
- Glue stick
- Magnet
- Chopstick or heavy-duty straw (optional)

THE STEPS

1. Design a background scene on one piece of cardboard that will be able to feature moving objects, animals, or imaginary creatures. The background should be able to stand its own by either folding back some of the cardboard or taping the cardboard up.
2. Decide on the moving parts of your scene. These could include things like cars, clouds, or silly monsters. Challenge yourself to make at least three moveable pieces.
3. Design your objects or characters on the other piece of cardboard. Cut them out with scissors, then use a glue stick to adhere a paper clip to the back of each one individually.

4. Once your background scene and characters are complete, hold the magnet behind the cardboard scene and use it to move the objects glued to the paper clips. Note: For a longer reach, you can attach the magnet to the end of a chopstick or heavy-duty straw.

Now Try This! Create multiple scenes. Use them as your inspiration to write a play or short story that can be acted out for an audience using multiple magnets.

The Hows and Whys: Some materials, like air, wood, cardboard, plastic, and brass, cannot support the formation of a magnetic field; these are said to have no permeability. This means that an outside magnetic field cannot attract them. Other materials, such as iron, nickel, and cobalt, have high permeability. This means that magnetic fields can be formed in them when exposed to an outside magnetic field. Because the cardboard allows magnetic-field lines to pass through it, but the metal in the paper clip is permeable, the magnet is able to attract the paper clip on the other side of the cardboard scene.

TEMPORARY MAGNETS

How can you turn everyday objects into temporary magnets?

LEVEL OF DIFFICULTY: EASY
TIME SUGGESTION: 10 MINUTES

MATERIALS

- Metal paper clips
- Paper plate
- Scissors with metal blades
- Bar magnet

THE STEPS

1. Lay a pile of paper clips on a paper plate. Touch the scissors to the pile of paperclips to confirm that the they are not magnetized.
2. With adult supervision, rub the bar magnet along the blades of the scissors several times in one direction.
3. Lower the scissors toward the pile of paperclips. Record your observations. Before repeating the experiment, you will need to drop the scissors on the floor or another hard surface a few times to undo the magnetic domain.

Now Try This! Try making temporary magnets out of other metal objects around your home.

The Hows and Whys: Temporary magnets show signs of magnetism only when exposed to strong magnetic fields. By rubbing the magnet in one direction down the scissor blade, you align the molecules (magnetic domain) in the scissors in such a way that one side is acting like the north end of the magnet while the other is acting like the south end, making them a temporary magnet. When you drop the scissors a few times, you jumble up the molecules again, causing the scissors to lose their magnetic properties.

Chapter Seven

Light is a form of energy that enables us to see the world around us. It is emitted from a source such as the Sun or a light bulb and then reflects off an object into your eye, allowing you to see the object. You already know that mirrors and shiny metals reflect light, but other materials reflect light, as well. The reason we can see trees, rocks, walls, books, and even the moon is because they reflect light. When no light is available to reflect off an object into your eye, you cannot see anything. This is why it's hard to read a street sign at night without headlights or read a book in bed in the dark.

Light can travel in a straight line, can reflect off objects, and can bend as it passes around or through something. For example, when a rainbow appears in the sky, you are actually seeing sunlight refracting and reflecting in water droplets.

In this chapter, you will beam with excitement as you find you are able to use light, along with a few common materials, to write messages in the air, see around corners, and create holograms. However, just as you do with any science investigation, you'll want to use caution. When working with the different sources of light, make sure you never direct them toward anyone's eyes or look directly at them yourself. What are you waiting for? C'mon and turn the page!

DISAPPEARING GLASS

How can you make an entire bowl disappear?

LEVEL OF DIFFICULTY: EASY

TIME SUGGESTION: 10 MINUTES

MATERIALS

- Large glass bowl
- Cooking oil
- Small Pyrex bowl
- Penny

THE STEPS

1. Fill a large glass bowl with cooking oil about two thirds full.
2. Slowly submerge the Pyrex bowl in the oil until it's completely covered. Slowly turn it upside down, making sure there are no air bubbles.
3. Have someone check out the bowl of oil. Can they see anything inside?
4. Drop a penny in the oil so that it lands on top of the upside-down bowl. Does it look like it's suspended?

Now Try This! Try different glass objects, such as marbles, test tubes, or glass jars, to see if they'll disappear, too.

The Hows and Whys: When light moves from one medium to another, it bends slightly and changes speed. This is why when you look at a pencil placed in a short glass of water, it appears to be broken—the section in the air seems to be separated from the section in the water. Under normal circumstances, this refraction, or bending of light, allows our brain to understand that we are looking at something transparent. However, light travels at the same speed through vegetable oil as it does through Pyrex glass, so no bending of the light occurs and our brain can't tell the difference between the two, making it look like the bowl has disappeared.

LIGHT BENDER

What makes light bend in a stream of water?

LEVEL OF DIFFICULTY: EASY

TIME SUGGESTION: 15 MINUTES

MATERIALS

- Duct tape
- Empty plastic water bottle, label removed
- Thumbtack
- Clear tape
- Water
- Tray (or pan) to catch water
- Laser pointer

THE STEPS

1. Place a small piece of duct tape halfway down an empty plastic water bottle. Use a thumbtack to poke a hole in the center of the duct tape.
2. Cover the hole with clear tape and fill the bottle with water.

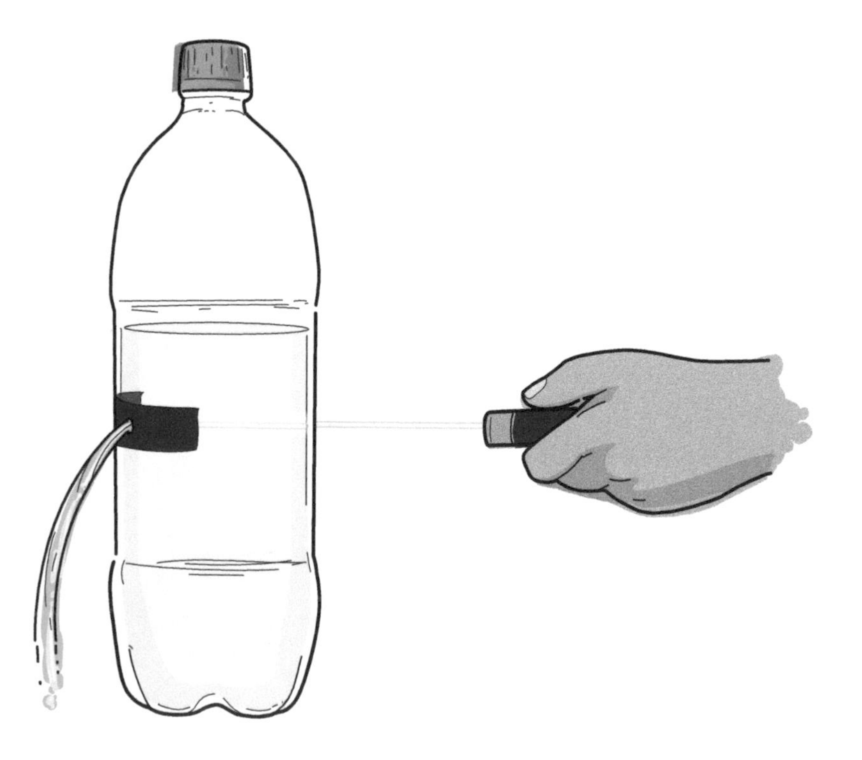

3. Place the bottle in the center of a tray. When ready, remove the clear tape to start a steady stream of water.
4. Shine a laser pointer directly at the hole from the opposite side of the bottle and watch how the beam of light comes out the other side. Record your observations.

Now Try This! Use a nail and make the hole a little larger. Does it make a difference?

The Hows and Whys: The light beam inside the water stream behaves as it would in an optical fiber. Optical fibers are how we transfer information, such as sound, pictures, or computer data, in the form of light. Just like the water stream, it doesn't matter if the fiber is straight or bent; the light beam travels through and comes out the other end because of a total internal reflection. Total internal reflection is when the light beam bounces around the inside of the fiber, or the water stream in this case, reflecting back and forth off the inside surfaces, and doesn't stop until it comes out the far end.

3D HOLOGRAMS

How is a hologram created?

LEVEL OF DIFFICULTY: MEDIUM

TIME SUGGESTION: 20 MINUTES

MATERIALS

- Plastic sheet, such as from media packaging, or hardened laminator sheet
- Marker
- Ruler
- Scissors
- Clear tape
- Smartphone (or tablet) with Internet access

THE STEPS

1. On a plastic sheet, trace the template on page 76 with a marker.
2. Have an adult help you use a ruler and the edge of a scissor blade to score the plastic where the dotted green lines are on the template so that it bends easily.
3. Tape the outside edges together so the final piece is in the shape of a pyramid. There will be a hole at the top. Make sure the edges are straight so that when the pyramid is flipped upside down, it can balance on a flat surface.
4. Ask an adult for permission to use a smartphone or tablet to search for a hologram video on the Internet—there are many! (See the Resources section on page 112.)

5. Set your hologram pyramid upside down on the smartphone or tablet, turn off the lights, and hit Play. Enjoy the show!

Now Try This! Can you make a larger hologram display?

The Hows and Whys: You may notice the images on the video are four identical versions of the same image. When each image is reflected off the four faces of the pyramid, the reflections angle toward the center. The projections, when all seen at the same time, create a 3D illusion for the viewer and makes it appear that the image is floating.

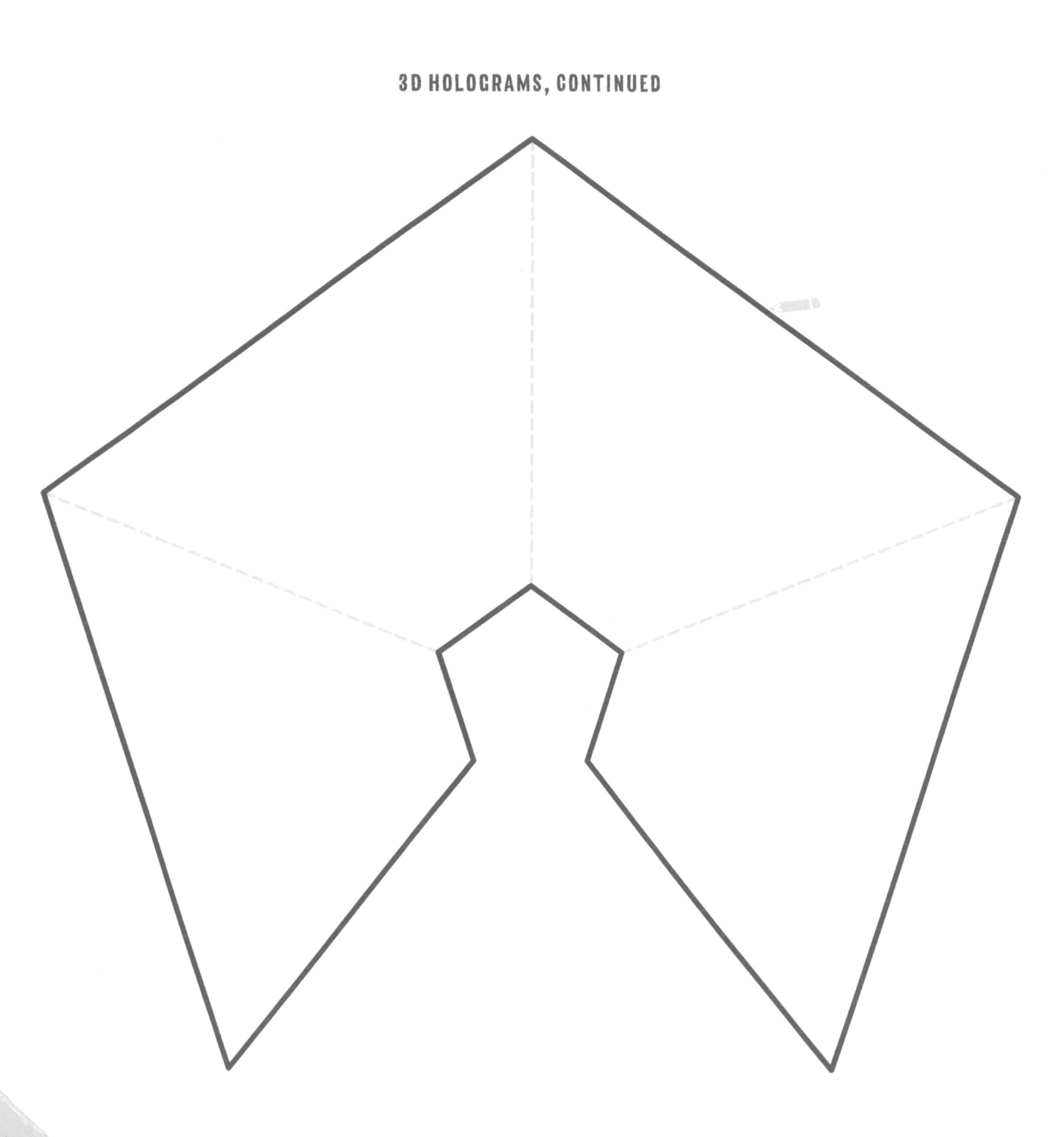

PHOTON-GRAPH

How can you capture a message written in light?

LEVEL OF DIFFICULTY: EASY

TIME SUGGESTION: 10 MINUTES

MATERIALS

- Electrical tape
- LED light
- Watch battery
- Digital camera (or device with camera and slow-shutter app installed)
- Tripod (optional)

THE STEPS

1. Use electrical tape to attach an LED to a watch battery. Make sure the longer leg of the LED is touching the positive end of the battery.
2. Place a digital camera on a solid surface or affix it to a tripod to make sure it doesn't move.
3. Set the shutter speed of the camera to very slow by manually adjusting the setting to $1/60$ or slower.
4. Darken the room and stand in front of the camera with your LED powered on.
5. Have someone press the camera button while you use the LED to draw a smiley face or write a short message in the air. Hint: For a word message, you will need to write backward from right to left, so you may need to practice a few times.

Now Try This! Try using both hands and different-colored LEDs to see what awesome designs and messages you can come up with. Adjust the shutter speed on the camera to see what setting gives you the best results.

The Hows and Whys: When you press the button on the camera, called the shutter release, it opens the shutter and an image is recorded. When the shutter closes, the camera stops recording. The shutter speed is how long the image is exposed to light. When the shutter is set to slow, it can capture the light of the LED over a longer period, recording the full image or message you created.

UP PERISCOPE!

How do light and reflections help you see around corners?

LEVEL OF DIFFICULTY: MEDIUM

TIME SUGGESTION: 30 MINUTES

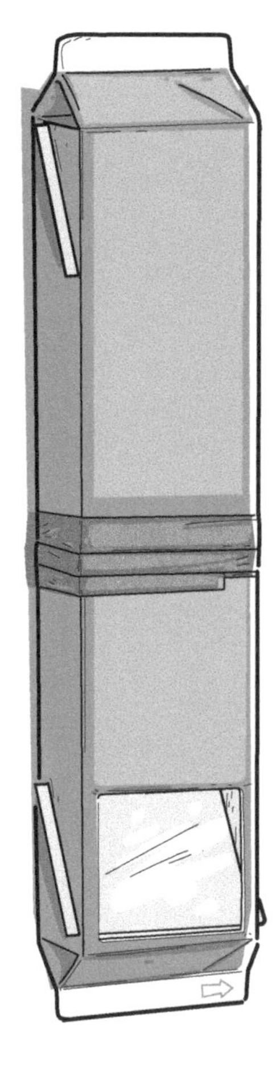

MATERIALS

- Scissors or craft knife
- 2 (1-quart) empty cardboard milk cartons, tops removed
- Marker
- 2 rectangular pocket mirrors (they should be same size)
- Masking tape
- Cardboard (if needed)

THE STEPS

1. With the help of an adult, use scissors to cut a rectangular window at the base of both milk cartons. Leave a 1- to 2-centimeter border on the sides and bottom.
2. Lay each carton on its side with the window facing right. On the side that's facing up, use a marker to draw a line at a 45-degree angle from the bottom right-hand corner to the opposite edge of each carton. Flip each carton over so that the window is now facing left, and on the side facing up, draw a line at a 45-degree angle from the bottom left-hand corner to the opposite edge.

3. Make a cut on each drawn line only as long as the shorter side of the mirrors.
4. Insert a mirror through the slits of each carton so that the reflecting side is facing the window. Tape in place if necessary. Note: If the mirrors are too small, you can tape them to pieces of cardboard, then slide the cardboard through the slits.
5. Place one carton on a table with the window facing you. Place the other carton on top, upside down, with the window facing away from you.
6. Slightly pinch in the top milk carton so that it slides inside the bottom carton. Use tape to secure it in place.
7. Use your periscope to peek around edges, over fences, or under tables. Go explore!

Now Try This! Try using longer tubes (you can add two milk cartons together), or make the periscope adjustable to see over higher walls. Is the image still the same if longer tubes are used?

The Hows and Whys: The angle at which light reflects off of a mirror is the same angle at which it hits the mirror—this is called the angle of reflection. In the periscope, light hits the top mirror at a 45-degree angle and reflects away at the same angle, which bounces it down. That reflected light hits the bottom mirror at the same angle, then reflects away, again at a 45-degree angle, right into your eye. You can now see the image that is around the corner without moving around the corner. Pretty sneaky!

Chapter Eight

HEAT

If you placed an ice cube in the palm of your hand, then made a fist around the ice cube for 10 seconds, what would happen? You may say that the ice cube would make your hand cold. However, I have a feeling that after you complete the activities in this chapter, you'll have a much better explanation for what's happening between your hand and the ice cube.

You've probably sat on a chair or couch at home and could tell if some-one had been sitting there recently. The reason for this is that the person's body heat flowed to the seat and increased its temperature. Heat, or thermal energy, always travels from warmer objects to cooler objects and is trans-ferred from one thing to another by conduction, convection, and radiation.

In this chapter, you will not only investigate each type of heat transfer, but you will also learn how to control the flow of heat using insulators. You probably don't even realize you are already controlling the flow of heat on a daily basis. For example, when you bake brownies in the oven, do you put on oven mitts to pull them out when they are done? Of course you do, because if you didn't, you would get seriously injured. So, now is a good time to remind you that when working with heat, especially flames and hot liquids, always ask an adult to supervise.

Are you fired up for some heat investigations? You should be—we are going to be cooking up s'more!

UNDERWATER CANDLE

How can you keep a candle burning under water?

LEVEL OF DIFFICULTY: EASY

TIME SUGGESTION: 10 MINUTES

MATERIALS

- Matches
- Taper candle, cut to the height of the bowl
- Clear bowl
- Cold water

THE STEPS

1. With the help of an adult, light a candle and let the wax drip in the center of a clear bowl until you have a wax circle the size of a nickel.
2. Blow out the candle and press it into the warm wax to hold it in place.
3. Carefully fill the bowl with cold water to just below the top of the candle. Make sure you do not get the wick wet.
4. With the help of an adult, light the candle and watch as the flame burns below the surface of the water!

Now Try This! Try different sizes of candles or different temperatures of water to see how the results change. Make sure to record what variable you're changing—and the different outcomes—in your science notebook.

The Hows and Whys: Water is an excellent heat absorber, as you'll learn with the Heat-Resistant Balloon (page 84). The cool water surrounding the burning candle absorbs the heat from the flame, allowing the outer layer of wax to stay solid. Since it's not melting, a hollow tube forms in the candle, and the wick can burn down below the water's surface.

HEAT-RESISTANT BALLOON

What can keep a balloon from popping when held in a flame?

LEVEL OF DIFFICULTY: EASY

TIME SUGGESTION: 10 MINUTES

MATERIALS

- 2 balloons
- Match
- Candle
- Water

THE STEPS

1. Blow up a balloon and secure it with a knot. With the help of an adult, use a match to light a candle.
2. Hold the balloon high above the flame, then move it closer to the flame until it pops.
3. Fill the second balloon with 2 to 3 tablespoons of water while it's deflated. Without spilling the water, blow up the balloon and secure it with a knot.
4. Again, hold the balloon high above the flame, then slowly move it closer to the flame. See if you can hold the balloon in the flame for 5 seconds. Record your observations.

Now Try This! Try making the water colder and see if it makes a difference in how long the balloon can stay in the flame.

The Hows and Whys: As a substance absorbs heat, its temperature changes depending on the nature of the substance as well as the amount of heat added. When you move the first balloon closer to the flame, the heat melts the latex, which causes the balloon to pop. When you heat the second balloon, however, the water absorbs the heat by acting as a coolant. As the water closest to the flame heats up, the warmer water will rise and cooler water will replace it. This will start a convection current and keep the balloon from popping.

LAYER OF INSULATION

How does insulation affect the flow of heat?

LEVEL OF DIFFICULTY: MEDIUM

TIME SUGGESTION: 30 MINUTES

MATERIALS

- Tape
- 3 (2½-by-1½-centimeter) strips of black construction paper
- 3 thermometers
- 2 fold-top plastic sandwich bags
- Gallon-size plastic bag

THE STEPS

1. Tape one strip of construction paper over the bulb of each thermometer. This will help them absorb heat from the Sun.
2. Place one thermometer inside one sandwich bag and blow air into the bag to inflate it. Secure the air in the bag by twisting the opening around the thermometer, then sealing it with a piece of tape.
3. Place the second thermometer in the other sandwich bag, and again, inflate and seal it. Place this bag inside a gallon-size plastic bag. Inflate the gallon-size bag with air and seal it.

4. Leave the third thermometer unbagged. Read the initial temperature on each thermometer and record the information in your science notebook.

5. Place all three thermometers in sunlight. After 10 minutes, read the temperature of each thermometer and record the information in your science notebook.

Now Try This! Replace the thermometers with ice cubes and see if the double-bag insulator helps keep the ice from melting.

The Hows and Whys: Many insulators contain pockets of trapped air. These air pockets conduct heat poorly and also keep convection currents from forming. Building insulation is usually made of fiberglass, which contains pockets of trapped air. The insulation is packed into outer walls and attics, where it reduces the flow of heat between the inside of the building and the outside environment. Another way builders make insulation work to their advantage is by installing double-paned windows that have trapped air between the two panes of glass, which helps keep the cold out during the winter, and the heat out during the summer.

HOT-AIR BALLOON

How can you inflate a balloon without blowing into it?

LEVEL OF DIFFICULTY: EASY

TIME SUGGESTION: 35 MINUTES

MATERIALS

- Balloon
- Empty 16.9-ounce plastic water bottle
- Hot water

THE STEPS

1. Blow up a balloon a few times so that it's stretched out in advance to get the best results. Place the balloon over the opening of an empty water bottle, making sure to pull it down over the mouth of the bottle.
2. Place the water bottle with the balloon in the freezer for 30 minutes.
3. With adult supervision, fill a sink, pitcher, or pot with very hot water.
4. Remove the bottle from the freezer. The balloon should be inside the bottle.
5. Immediately submerge the water bottle in the hot water until the water is about three quarters up the side of the bottle. Hold it in place. Record your observations.

Now Try This! Use a 2-liter bottle and see if the results are more dramatic. You can also try reversing the experiment by placing the bottle with the balloon in hot water first, then placing it in the freezer.

The Hows and Whys: When the bottle is placed in the freezer, the air molecules in the bottle condense, causing the pressure to decrease, which makes the balloon get sucked inside the bottle. When the bottle is then placed in hot water, the heated air inside starts moving faster, causing the molecules to move farther apart. This also increases the pressure, causing the balloon to inflate.

SOLAR-POWERED SNACK

Can you cook s'mores using energy from the Sun?

LEVEL OF DIFFICULTY: DIFFICULT

TIME SUGGESTION: 2 HOURS

MATERIALS

- Empty pizza box
- Scissors
- Aluminum foil
- Glue
- Black construction paper
- Tape
- Plastic wrap
- Wooden stick
- Paper plate
- Graham crackers
- Marshmallows
- Chocolate bars

THE STEPS

1. With the pizza box closed, carefully cut a three-sided flap in the lid using scissors, leaving a 1- to 2-inch border around the flap. Glue a piece of aluminum foil to the inside of the flap.
2. Open the lid and line the rest of the inside of the box with aluminum foil, using glue to hold it in place. Be sure to make the foil as smooth as possible. The more mirror-like it is, the better your solar oven will work.
3. Glue a sheet of black construction paper to the center of the bottom of the box.
4. With the flap open, tape a layer of plastic wrap across the flap opening in the lid. Open the lid and tape another layer of plastic wrap across the opening, this time from the bottom. You should have two layers of plastic covering the opening, underneath the flap.
5. Prop open the lid with the wooden stick. You may want to make an indent or a small hole to hold the stick in place so it does not slide. Set the oven in direct sunlight to preheat for 15 to 20 minutes.
6. While you're waiting for the solar oven to preheat, line a paper plate with aluminum foil.
7. Place two graham cracker halves on the plate and place a marshmallow on each.

8. Place the plate inside the solar oven, close the lid, and prop open the flap. How hot it is outside will determine how long it takes for your marshmallows to get soft. This works best if the outside temperature is above 80°F.

9. Once the marshmallows are soft, add a piece of chocolate on each and top with another graham cracker half. Enjoy!

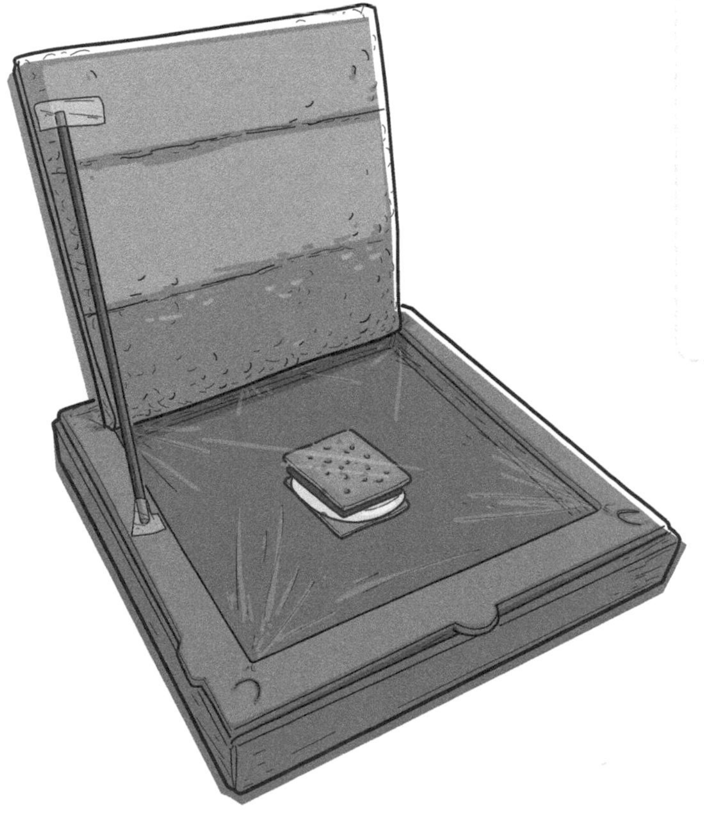

Now Try This! Research solar ovens and see how you can make changes to your design to make a more efficient cooker. What other tasty treats can you cook up in your solar oven?

The Hows and Whys: Solar ovens use solar energy—light and heat emitted from the Sun—which cooks food by reflecting off the foil in the box. Using full, direct sunlight will give the best results. The plastic window you added acts as an insulator by allowing sunlight to pass into the box but also keeping heat inside. The black paper on the bottom warms up by absorbing direct and reflected sunlight, helping make your gooey, warm treat.

Chapter Nine

If I asked you to imagine the sounds of an amusement park, which ones would you think of? Amusement parks can be noisy places, with the screaming of riders on roller coasters, carousel music tinging away, and bells and whistles from games and booths. The reason you can differentiate all these sounds is that they each have different pitch, loudness, and amplitude. However, they also have something in common: Each sound is produced by an object that vibrates. This property of sound is very important to understand because it's the basic principle behind the following activities on sound.

Although we can't see or touch sound, we still have to use caution when doing physics experiments with sound. The reason for this is that all those vibrating objects create sound waves, which are gathered by the outer part of the ear and sent to the eardrum. Once inside the ear, those sound vibrations travel through the middle of it, where tiny bones multiply the force and pressure of the sound wave, amplifying the sound. The amplified sound hits 20,000 to 30,000 tiny hair cells in the inner ear. When vibrations reach those hair cells, electrical impulses send messages to the auditory center of the brain, and we are able to identify the vibrations as music, pots clanging, or a car driving past.

It's important to note how the ear works, because our ears are not meant to withstand very loud sounds. If a sound is too loud, like when headphones are set at an extremely high volume, the hair cells are damaged and may not grow back, resulting in hearing loss. While the following activities will not produce damaging vibrations, be careful if you dive into future sound investigations that require ear protection.

Ready to investigate some good vibrations? I'm sure your family will be all ears when you share what you have done.

PHONE SPEAKERS

How can you make your phone louder if it's already turned all the way up?

LEVEL OF DIFFICULTY: EASY

TIME SUGGESTION: 20 MINUTES

MATERIALS

- Smartphone or music-playing device with some good songs
- Sturdy gift wrap tube, cut to about 25 centimeters long
- Marker
- Box cutter
- 2 plastic cups
- Hot glue gun

THE STEPS

1. Stand the phone on the middle of the tube and carefully trace its width and depth using a marker. With the help of an adult, use the box cutter to carefully cut out a slot that the phone can slide into.
2. Trace one end of the tube onto one side of each cup. Carefully cut out each circle with the box cutter.
3. Slide a cup onto each end of the tube with the tops of the cups facing you. Adjust the position of the tube so that the phone slot is almost directly on the top but slightly angled toward the back. Use the hot glue gun to secure the cups to the tube.
4. Start playing your tunes, then insert your phone into the slot. Let the dance party begin!

Now Try This! Use different lengths of tube or use paper cups to see what combination gives your music the best amplified sound.

The Hows and Whys: Music is sound being sent to your ear by waves. Usually when the sound comes from your phone speaker, the waves move out in all directions away from the phone. When you place the phone inside the tube, the sound waves vibrate down the tube into the cups, which reflect the sound waves that would normally go away from you toward you, making the sound amplified in the front and not as loud in back.

BUZZING BEE FLYER

How can you produce a sound like a swarm of bees?

LEVEL OF DIFFICULTY: EASY

TIME SUGGESTION: 20 MINUTES

MATERIALS

- Play-Doh
- Large craft stick
- Scissors
- Index card
- Duct tape
- 60-centimeter-long string
- Wide rubber band long enough to stretch the length of the craft stick

THE STEPS

1. Attach a marble-size piece of Play-Doh to each end of a craft stick.
2. Use scissors to trim an index card so it fits between the Play-Doh pieces. Tape it in place with duct tape.
3. Tie the string to one end of the craft stick as close as possible to the Play-Doh.
4. Stretch the rubber band around the craft stick lengthwise and make sure it's secure in the Play-Doh. Be sure that the knot in your string is not touching the rubber band.
5. Find a clear area and swing your buzzing bee flyer overhead. Record your observations.

Now Try This! Cut the index card into different shapes or use a thicker piece of paper to see if there is a difference in the sound you hear. Try changing the length of the string, which can also make you swing the flyer faster or slower. Now how does the sound change?

The Hows and Whys: As the flyer moves through the air, the rubber band begins to vibrate. The sound produced by the vibration is amplified by the index card, sounding like a swarm of bees.

PHONOGRAPH

How does a record produce sound?

LEVEL OF DIFFICULTY: EASY

TIME SUGGESTION: 15 MINUTES

MATERIALS

- Paper
- Tape
- Straight pin
- Record (make sure you have permission to use it)
- Record player

THE STEPS

1. Take a piece of paper and, starting in one corner, roll it up to form a cone. The pointy end of the cone should be closed. Tape the edge of the cone to secure it and keep it from unrolling.
2. Push the straight pin at an angle through the cone about 1 centimeter from the tip. Make sure the pin is through both sides of the cone and points down at roughly a 45-degree angle.
3. Tape the head of the pin to the cone to secure it in place.
4. Place the record on the record player and turn it on.
5. Lightly hold the outer edge of the cone so that the pin is pointed down. Rest the pin in the grooves of the record and listen. Record your observations of what you hear.

Now Try This! Try making different-size cones or use different materials to make a cone and see if there are changes in the results.

The Hows and Whys: Records have undulations, or bumps and bends, in the grooves that correspond to a specific song. When the pin travels along the groove, it vibrates back and forth in accordance with these bumps. The vibrations of the needle are amplified, or increased, by the cone and travel to your ear, which your brain translates as music.

UPCYCLED WIND CHIME CHALLENGE

Can you design and build a wind chime that plays four different notes using discarded items?

LEVEL OF DIFFICULTY: MEDIUM

TIME SUGGESTION: 1 HOUR 30 MINUTES

MATERIALS

- Discarded items that can chime, such as old silverware, old keys, empty soup cans, shells, PVC pipe scraps, and metal bottle caps
- Drill (optional)
- Materials to decorate your wind chime, such as paint or markers
- Scissors
- String
- Ruler
- Tape
- Hot glue gun
- Embroidery hoop (or other circular hoop)
- Fan

THE STEPS

1. Research parts of a classic wind chime, noting different lengths of tubes, how materials are suspended, and spacing of objects (see Resources, page 112). Next, gather materials from around your home that could be used for an upcycled wind chime. Have an adult drill a hole in the objects that don't already have one.
2. Decorate the parts of your wind chime with paint, markers, or anything crafty to make them look colorful and worthy of hanging outside your home.
3. Use scissors, string, a ruler, tape, and/or a hot glue gun to construct your wind chime so that all the parts are securely fastened to an embroidery hoop. Test your wind chime in front of a fan to make sure that it operates continuously without getting tangled and makes sounds pleasing to the listener.
4. Hang it for all to enjoy!

CONTINUED

Now Try This! Evaluate your wind chime and record all the dimensions, such as length of objects, object placement, and so on. How is the quality of sound? Can you modify, or make changes to, your original design for a different sound? You can even add a sail, which will catch the wind to help move the wind chime.

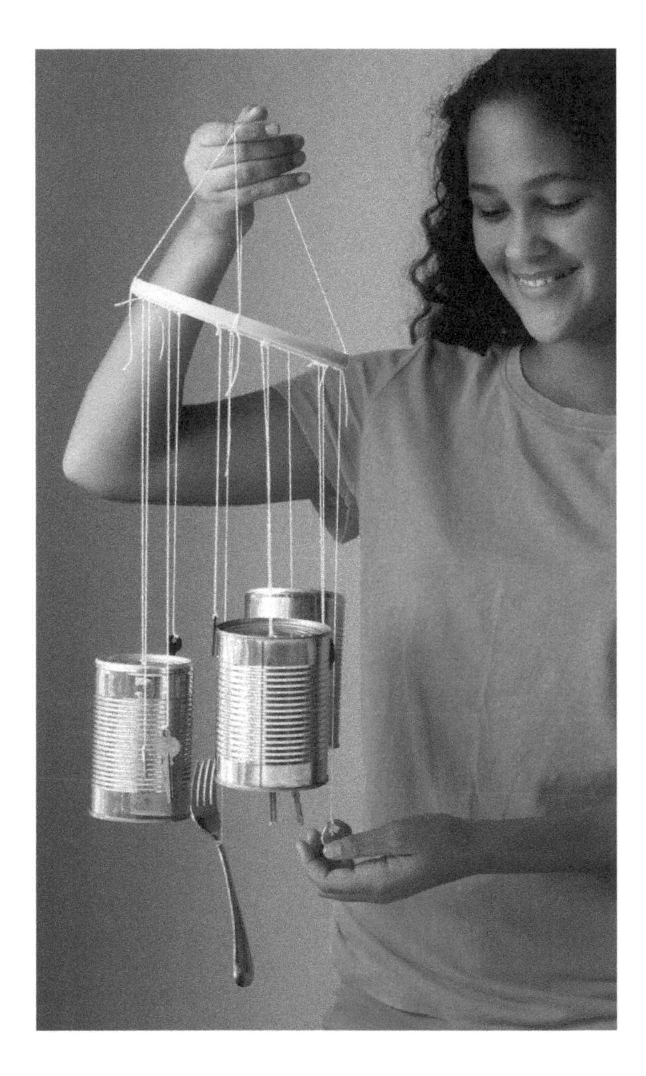

The Hows and Whys: A wind chime usually makes its signature sound because of the metal tubes that are suspended from the top. The lengths of the tubes determine if high-pitched or low-pitched sounds are produced. Shorter tubes produce higher tones than longer tubes. Other wind chimes mix elements of metal and wood to produce different sounds that appeal to different listeners. No matter which material you use, wind is the driving force that makes your wind chime components collide to produce sound.

MAKE IT RAIN

What causes the sound of falling rain in a rain stick?

LEVEL OF DIFFICULTY: MEDIUM

TIME SUGGESTION: 30 MINUTES

MATERIALS

- Scissors
- Ruler
- Aluminum foil
- Pencil
- Long cardboard tube, such as a wrapping paper tube
- 2 index cards
- Masking tape
- 1 cup uncooked rice

THE STEPS

1. Using scissors and a ruler, cut sheets of aluminum foil into 10-centimeter strips so when they are laid end to end, you have enough to equal a length and a half of a cardboard tube.
2. Attach the end of one aluminum foil strip to the end of another by loosely scrunching them together, making sort of a foil snake. Repeat until you have one long, continuous aluminum foil snake.
3. Starting at one end, wrap the foil around a pencil to form a spiral. Continue until the entire length is coiled. Insert the coiled aluminum foil into the cardboard tube. Secure the foil to each end of the tube with masking tape.

4. Cut the index cards large enough to cover the tube openings. Secure one index card with tape to one end of the tube.
5. Pour the rice into the tube through the open end. Cover the other end of the tube with the second index card and secure it with tape.
6. Tip the tube upside down and listen to the sound until all the rice has fallen.

Now Try This! Instead of rice, try using other materials, such as dried pinto beans or sesame seeds, to see if they produce a different sound. Also, try changing the length of the tube to see if it makes a difference in sound.

The Hows and Whys: When a grain of rice bumps into the aluminum foil as it falls down the tube, the collision causes a vibration from the foil to the cardboard tube, sending the sound through the air, which our ears hear as a soft click. When many pieces of rice fall at once, the clicks happen randomly, producing a white-noise sound similar to a rain shower.

Chapter Ten

PUTTING IT ALL TOGETHER

Aw, man. Is it over already? Take a look back at all the amazing physics activities and investigations you've completed. I can imagine you trying to decide which investigation to do again. That's because physics is amazing, as you have seen through these fun, hands-on experiments.

Over the course of this book, you have not only learned some fundamentals of physics, but you've also learned how to be an excellent scientist. How do I know? Well, throughout the activities, you've made many observations, you've collected data, you've done error analysis when things haven't gone quite right, and you've probably even changed some variables to see if they produced different results. These are all skills that scientists use to learn about what they are studying—regardless of the level at which they are practicing. You have conducted experiments in the same way that scientists who are changing the world have. If you continue your studies, you could change the world, too.

Didn't you build a circuit and even make a flashlight out of pennies? When you apply science to your everyday life and make things easier, you are working with technology. Technology cannot be avoided in our day-to-day lives and will continue to quickly advance over the years. Understanding the basic physics principles behind technology will springboard you to becoming a great innovator.

By successfully working through the activities in this book, you were able to solve problems, think critically, and pay attention to detail. These are common skills of a great engineer. Engineers design, test, modify, construct, and conduct error analysis. Yep, I'm pretty sure you've hit all those, but what's more, by combining those skills with the physics you've learned, you've actually had a little taste of many different engineering fields. For example, when you made and tested the rotors from chapter 3, Buoyancy and Flight, you were getting a sample of what an aerospace engineer works on. Electrical engineers design, develop, and manage electrical and electronic equipment. Check. How about mechanical engineers, who study motion, energy, and force to develop various solutions for mechanical systems? Double check. Are you starting to get the picture?

Now that you're done with the book, have you realized all the math you did? By measuring, calculating, combining data, and designing structures, your math skills were put to work, which made you a better problem solver and critical thinker.

You may be onto me by now about where I'm going with all of this, but if you haven't seen the pattern yet, your journey through these awesome physics experiments has tied together science, technology, engineering, and mathematics, or what is referred to as STEM.

STEM is not new; it's a description of how things work in the real world. What I tried to show in this book is that physics isn't best taught separately from engineering, technology, or even math; physics can be taught in a way that shows how the fields complement and support one another, just like in real life.

I want you to do one more thing for me that all great physicists do: Make a promise to yourself that you'll always be curious, ask questions, explore, and play. Do this, and you will be amazed at what you discover in this world. Your passion and love of learning will ultimately help you succeed at whatever lies ahead. Now, go keep that promise!

Glossary

absorb: To take in

acceleration: Rate of change of velocity

accuracy: Precision or exactness

adhesion: Attraction of two unlike things that hold together

amplify: To make larger, greater, or louder

amplitude: Measure of energy carried by a wave

angle of reflection: The angle at which a light wave reflects off a surface

atom: Smallest particle of an element that still retains property of the element

attract: To cause to move toward

buoyancy: The ability of a fluid to push up on an object immersed in it

centripetal force: Force directed toward the center of a curved or circular path

chemical reaction: Process in which one or more substances are changed into new substances

circuit: Closed loop through which an electric current can flow

cohesion: Attraction that holds two similar molecules together

colloid: Substance that shares properties of a liquid, solid, or gas

condensation: The changing of a gas into a liquid

conduction: Transfer of thermal energy by touch

convection: Transfer of thermal energy by the movement of fluid

density: Mass per unit of volume

displacement: Distance and direction of an object's change in position from the starting point

electric current: Movement of electric charges in a single direction

electrode: A conductor through which a current enter or leaves

electrolyte: Compound that breaks apart in water, forming charged particles that can conduct electricity

electromagnet: Temporary magnet made by wrapping a wire coil, carrying a current, around an iron core

electrons: Particles surrounding the center of an atom that have a negative charge

experiment: Organized procedure for testing a hypothesis

fluid: A substance that is capable of flowing, such as liquid or gas

force: Push or pull exerted on an object

friction: Force that opposes the sliding motion between two touching surfaces

gravity: Pulling force exerted by Earth

heat: Thermal energy that flows from a warmer material to a cooler material

hologram: A complete, three-dimensional photographic image of an object

hypothesis: Educated guess using what you know and what you observe

implosion: Sudden collapse

inertia: The tendency of an object to resist any change in its motion

insulator: Material in which heat flows slowly

kinetic energy: Energy a moving object has because of its motion

liquid: State of matter with a definite volume but without a definite shape

loudness: Human perception of sound intensity

magnetic domain: Group of atoms in a magnetic material with the magnetic poles of the atoms pointing in the same direction

magnetic field: The area that surrounds a magnet and exerts a force on other magnets and magnetic materials

mass: Amount of matter in an object

microwelds: Force that causes resistance when two surfaces try to slide over each other

momentum: Property of a moving object that equals its mass times its velocity

neodymium: A rare-earth metal used to make the most powerful permanent magnets in the world.

optical fiber: Very thin, flexible glass or plastic strand through which large quantities of information can be transmitted in the form of light

pendulum: An object suspended on a string from a pivot

permeability: The ability to support the formation of a magnetic field

photon: A fundamental particle of visible light

pitch: How high or low a sound seems

pivot: Fixed point

polymer: Long chain of molecules made from many smaller units connected together

potential energy: Stored energy an object has because of its position

pressure: Amount of force exerted per unit of air

projectile: An object thrown or shot forward through the air

property: Characteristic of a material that you can observe or attempt to observe

protons: Particles inside the nucleus of an atom that have a positive charge

radiation: Transfer of thermal energy by electromagnetic waves

reflection: Bouncing of light or a wave from a surface

refraction: Bending of light or a wave as it changes speed from one medium to another

repel: To cause to move away from

solid: State of matter in which material has a definite volume and shape

speed: The distance an object travels per unit of time

static electricity: Accumulation of excess electric charges on an object

surface tension: The elastic-like force existing on the surface of water due to the makeup of water molecules

transparent: Transmits almost all light so as to allow you to see through

undulation: Wavy form or outline

upcycle: To process used or waste materials so as to produce something better than the original

variable: Factor that can cause a change in the results of an experiment

velocity: Speed and direction of a moving object

volume: Space that an object takes up

water vapor: Water in the form of a gas

Measurement Conversions

Volume Equivalents (Liquid)

US Standard	US Standard (ounces)	Metric (approximate)
2 tablespoons	1 fl. oz.	30 mL
¼ cup	2 fl. oz.	60 mL
½ cup	4 fl. oz.	120 mL
1 cup	8 fl. oz.	240 mL
1½ cups	12 fl. oz.	355 mL
2 cups or 1 pint	16 fl. oz.	475 mL
4 cups or 1 quart	32 fl. oz.	1 L
1 gallon	128 fl. oz.	4 L

Oven Temperatures

Fahrenheit (F)	Celsius (C) (approximate)
250°F	120°C
300°F	150°C
325°F	165°C
350°F	180°C
375°F	190°C
400°F	200°C
425°F	220°C
450°F	230°C

Volume Equivalents (Dry)

US Standard	Metric (approximate)
⅛ teaspoon	0.5 mL
¼ teaspoon	1 mL
½ teaspoon	2 mL
¾ teaspoon	4 mL
1 teaspoon	5 mL
1 tablespoon	15 mL
¼ cup	59 mL
⅓ cup	79 mL
½ cup	118 mL
⅔ cup	156 mL
¾ cup	177 mL
1 cup	235 mL
2 cups or 1 pint	475 mL
3 cups	700 mL
4 cups or 1 quart	1 L
½ gallon	2 L
1 gallon	4 L

Weight Equivalents

US Standard	Metric (approximate)
½ ounce	15 g
1 ounce	30 g
2 ounces	60 g
4 ounces	115 g
8 ounces	225 g
12 ounces	340 g
16 ounces or 1 pound	455 g

Resources

RESOURCES FOR KIDS

nittygrittyscience.com
Lesson plans and resources geared toward physical science, earth science, and life science.

www.gardenhomey.com/wind-chime-parts-explained
The parts of a wind chime explained to help with the Upcycled Wind Chime Challenge (page 99).

www.exploratorium.edu
Videos and instructions for hands-on science activities you can make at home with household items.

www.stevespanglerscience.com/lab/experiments
Videos and tutorials for fun, hands-on science experiments.

www.youtube.com/hooplakidzlab
Videos that focus on everything science for kids! The creator has fantastic videos that show amazing experiments, many you can do at home.

www.sciencebuddies.org
Free project ideas and help in all areas of science, from physics to food science and music to microbiology.

physicscentral.com/experiment/colormephysics/index.cfm
A resource with free downloadable coloring pages and activities to introduce students to physics and some of the most famous physicists. There is even a coloring book of intricate and interesting physics images from physics journals and other sources.

www.nasa.gov/offices/education/centers/marshall/k-12/index.html
Information on unique opportunities for K–12 education in the Marshall Space Flight Center region and nationwide. Learn about ways to interact with NASA engineers and scientists in the programs listed on the website.

girlstart.org
A site focused on empowering girls to get involved with STEM through programs, hands-on activities, and volunteer opportunities.

RESOURCES FOR TEACHERS AND OTHER GROWN-UPS

www.edgalaxy.com/journal /2012/8/3/50-physics-lesson-plans -for-middle-school-students.html
A compiled list of 50 physics lessons plans geared toward middle school students.

www.instructables.com/teachers
Open-source website that has categorized a page for teachers and classrooms with integrated projects for design, electronics, and fabrication.

www.rubegoldberg.com /education/#free-lesson-plans
Lesson plans and resources to help you get started learning about Rube Goldberg machines for use in the classroom, homeschool, or youth group. Get a group of friends together and enter the yearly Rube Goldberg contest!

www.makefuncreating.com /categories/all-projects
A fun resource that offers easy to challenging projects, many using simple physics principles. Using the author's unique approach and downloads, you'll be building spaghetti towers, steampunk-inspired balancing robots, and a cardboard fort with a secret escape chute!

Blank Table

Blank Graph

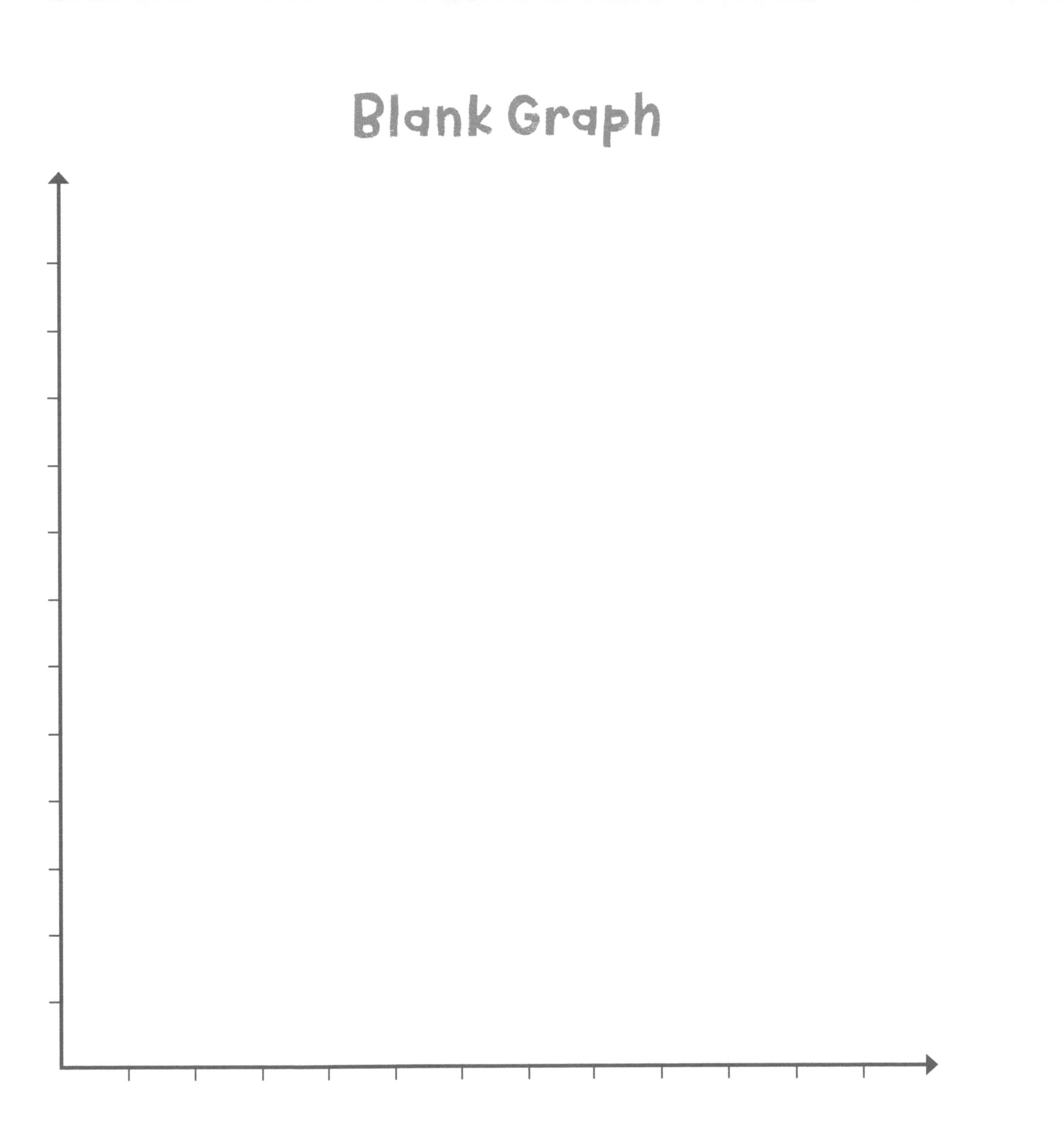

Index

R

S

T

U

Acknowledgments

We all have a great support team behind us, and this book wouldn't have been possible without mine.

First, thank you to my best friend and husband, Dan, who challenges me to be the best person I can be and who believes that I can do anything I set out to. Who knew life would twist and turn the way it has? But I'm so excited to be on this crazy journey with you.

To my girls, Ava and Dani, thank you for being just as excited as I am to do experiments. I hope you never lose your love of science and adventure. More important, I hope you never stop asking, "Why?" To my baby, Lincoln, thanks for being an easy one. I hope to guide you on your greatest adventure as you grow up.

Thanks to all my family and friends who have supported me and celebrated many milestones, including finishing this book! I feel like the grape industry should name some red blend after us or something. Cheers!

A special thanks to my parents. Mom, it all started with an article; thanks for thinking of me when you saw it. Dad, you've always been my biggest fan, and you've really asked for only one thing, so here it is: "Hi, Dad!"

Thank you to my editor, Brian Hurley, and Elizabeth Castoria for giving me such an awesome opportunity to work with you and your amazing team at Callisto Media. You have all been one of the most positive groups I've ever worked with. Thanks for making a bucket-list dream come true!

Finally, to all of the readers, thanks for sharing my passion for science. I hope you had as much fun with this book as I have!

About the Author

Dr. Erica L. Colón, PhD is a national board-certified teacher who taught science in the classroom for more than a dozen years. She earned her doctorate in Curriculum and Instruction and began to work with preservice science teachers. Over time, she would consistently hear from science teachers that all of their time was taken up trying to piece together resources to keep students engaged, as well as rigorous. Dr. Colón decided to design her own science curriculum to help teachers spend less time trying to make their own resources and focus on what they love to do most—teach. In 2012, she started her own business, Nitty Gritty Science. Her science curriculum can be found in classrooms all over the world. Dr. Colón is a proud Navy wife who currently lives with her husband and three children in Virginia.

CPSIA information can be obtained
at www.ICGtesting.com
Printed in the USA
JSHW011936281120
9822JS00002B/3